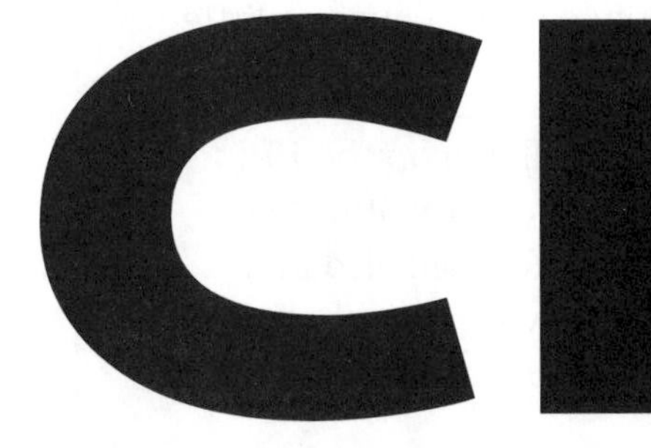

Discipline-Specific Review for the FE/EIT Exam

Robert H. Kim, MSCE, PE
with Michael R. Lindeburg, PE

Professional Publications, Inc.
Belmont, California

How to Locate Errata and Other Updates for This Book

At Professional Publications, we do our best to bring you error-free books. But when errors do occur, we want to make sure that you know about them so they cause as little confusion as possible.

A current list of known errata and other updates for this book is available on the PPI website at **www.ppi2pass.com**. From the website home page, click on "Errata." We update the errata page as often as necessary, so check in regularly. You will also find instructions for submitting suspected errata. We are grateful to every reader who takes the time to help us improve the quality of our books by pointing out an error.

CIVIL DISCIPLINE-SPECIFIC REVIEW FOR THE FE/EIT EXAM

Current printing of this edition: 7

Printing History

edition number	printing number	update
1	5	Minor corrections.
1	6	Updated front matter.
1	7	Updated front matter.

Printed in the United States of America

Professional Publications, Inc.
1250 Fifth Avenue, Belmont, CA 94002
(650) 593-9119
www.ppi2pass.com

Library of Congress Cataloging-in-Publication Data
Kim, Robert H., 1964–
Civil discipline-specific review for the FE/EIT exam / Robert H. Kim, with Michael R. Lindeburg.
p. cm.
ISBN 1-888577-18-5
1. Engineering--United States--Examinations--Study guides.
2. Civil engineering--United States--Examinations--Study guides.
3. Engineering--Problems, exercies, etc. 4. Engineers--Certification--United States. I. Lindeburg, Michael R.
II. Title.
TA159.K48 1997
624'.076--dc21 97-28080
CIP

Table of Contents

Preface and Acknowledgments

This book is one in a series of five that is intended for engineers and students who are taking the engineering discipline-specific (DS) afternoon session of the Fundamentals of Engineering (FE) exam.

The topics covered in the DS afternoon FE exams are completely different from the topics covered in the morning session of the FE exam. Since this book only covers one discipline-specific exam, it really addresses only half of the FE exam, and even then, only one specific discipline.

This book is intended to be a quick review of the material unique to the afternoon session of the civil engineering exam. The material presented covers the subjects most likely to be on the exam. This book is not a thorough treatment of the exam topics. Its objective is to prepare you with enough knowledge to pass. As much as was practical, this book uses the notation in the NCEES Handbook.

This book consolidates 120 practical review problems, covering all of the discipline-specific exam topics. The practice problems include full solutions. The topics are presented in essentially the same sequence followed by the NCEES Handbook, the only reference book permitted in the exam.

The problems in this book were developed by Robert H. Kim, MSCE, PE, following the format, style, subject breakdown, and guidelines that I provided.

In designing this book, I used the NCEES Handbook and the breakdown of problem types published by NCEES. However, as with most standardized tests, there is no guarantee that any specific problem type will be encountered. It is expected that minor variations in problem content will occur from exam to exam.

As with all of Professional Publications' books, the problems in this book are original and have been ethically derived. Although examinee feedback was used to determine its content, this book contains problems that are only *like* those that are on the exam. There are no actual exam problems in this book.

This book was designed to complement my *FE Review Manual*, which you will also need to prepare for the FE exam. The *FE Review Manual* is Professional Publications' most popular study guide for both the morning and afternoon general exams. It and the *Engineer-In-Training Reference Manual* have been the most popular review books for this exam for more than 20 years.

You cannot prepare adequately without your own copy of the NCEES Handbook. This document contains the data and formulas that you will need to solve both the general and the discipline-specific problems. A good way to become familiar with it is to look up the information, formulas, and data that you need while trying to work practice problems.

No exam-prep book is ever complete. By necessity, it will change as the exam changes. Even when the exam format doesn't change for a while, new problems and improved explanations can always be added.

Michael Lindeburg, PE
mlindeburg@ppi2pass.com

Engineering Registration in the United States

ENGINEERING REGISTRATION

Engineering registration (also known as *engineering licensing*) in the United States is an examination process by which a state's board of engineering licensing (i.e., registration board) determines and certifies that you have achieved a minimum level of competence. This process protects the public by preventing unqualified individuals from offering engineering services.

Most engineers do not need to be registered. In particular, most engineers who work for companies that design and manufacture products are exempt from the licensing requirement. This is known as the *industrial exemption.* Nevertheless, there are many good reasons for registering. For example, you cannot offer consulting engineering design services in any state unless you are registered in that state. Even within a product-oriented corporation, however, you may find that employment, advancement, or managerial positions are limited to registered engineers.

Once you have met the registration requirements, you will be allowed to use the titles Professional Engineer (PE), Registered Engineer (RE), and Consulting Engineer (CE).

Although the registration process is similar in all 50 states, each state has its own registration law. Unless you offer consulting engineering services in more than one state, however, you will not need to register in other states.

The U.S. Registration Procedure

The registration procedure is similar in most states. You will take two eight-hour written examinations. The first is the *Fundamentals of Engineering Examination*, also known as the *Engineer-In-Training Examination* and the *Intern Engineer Exam.* The initials FE, EIT, and IE are also used. This examination covers basic subjects from all of the mathematics, physics, chemistry, and engineering classes you took during your first four university years.

In rare cases, you may be allowed to skip this first examination. However, the actual details of registration qualifications, experience requirements, minimum education levels, fees, oral interviews, and examination schedules vary from state to state. Contact your state's registration board for more information.

The second eight-hour examination is the *Principles & Practice of Engineering Examination.* The initials PE are also used. This examination covers subjects only from your areas of specialty.

National Council of Examiners for Engineering and Surveying

The National Council of Examiners for Engineering and Surveying (NCEES) in Clemson, South Carolina, produces, distributes, and scores the national FE and PE examinations. The individual states purchase the examinations from NCEES and administer them themselves. NCEES does not distribute applications to take the examinations, administer the examinations or appeals, or notify you of the results. These tasks are all performed by the states.

Reciprocity Among States

With minor exceptions, having a license from one state will not permit you to practice engineering in another state. You must have a professional engineering license from each state in which you work. For most engineers, this is not a problem, but for some, it is. Luckily, it is not too difficult to get a license from every state you work in once you have a license from one state.

All states use the NCEES examinations. If you take and pass the FE or PE examination in one state, your certificate will be honored by all of the other states. Although there may be other special requirements imposed by a state, it will not be necessary to retake the FE and PE examinations. The issuance of an engineering license based on another state's license is known as *reciprocity* or *comity.*

The simultaneous administration of identical examinations in all states has led to the term *uniform examination.* However, each state is still free to choose its own minimum passing score and to add special questions and requirements to the examination process. Therefore, the use of a uniform examination has not, by itself, ensured reciprocity among states.

THE FE EXAMINATION

Applying for the Examination

Each state charges different fees, specifies different requirements, and uses different forms to apply for the exam. Therefore, it will be necessary to request an application from the state in which you want to become registered. Generally, it is sufficient for you to phone for this application. You'll find contact information (websites, telephone numbers, email addresses, etc.) for all U.S. state and territorial boards of registration at www.ppi2pass.com. Click on the State Boards link.

Keep a copy of your examination application and send the original application by certified mail, requesting a receipt of delivery. Keep your proof of mailing and delivery with your copy of the application.

Examination Dates

The national FE and PE examinations are administered twice a year (usually in mid-April and late October), on the same weekends in all states. Check www.ppi2pass.com for a current exam schedule. Click on the Exam FAQs link.

FE Examination Format

The NCEES Fundamentals of Engineering examination has the following format and characteristics.

- There are two four-hour sessions separated by a one-hour lunch.
- Examination questions are distributed in a bound examination booklet. A different examination booklet is used for each of these two sessions.
- The morning session (also known as the *A.M. session*) has 120 multiple-choice questions, each with four possible answers lettered (A) through (D). Responses must be recorded with a pencil provided by NCEES on special answer sheets. No credit is given for answers recorded in ink.
- Each problem in the morning session is worth one point. The total score possible in the morning is 120 points. Guessing is valid; no points are subtracted for incorrect answers.
- There are questions on the examination from most of the undergraduate engineering degree program subjects. Questions from the same subject are all grouped together, and the subjects are labeled. The percentages of questions for each subject in the morning session are given in the following table.

Morning FE Exam Subjects

subject	percentage of questions (%)
chemistry	9
computers	6
dynamics	7
electrical circuits	10
engineering economics	4
ethics	4
fluid mechanics	7
materials science and structure of matter	7
mathematics	20
mechanics of materials	7
statics	10
thermodynamics	9

- There are seven different versions of the afternoon session (also known as the *P.M. session*), six of which correspond to a specific engineering discipline: chemical, civil, electrical, environmental, industrial, and mechanical engineering.

  Each version of the afternoon session consists of 60 questions. All questions are mandatory. Questions in each subject may be grouped into related problem sets containing between two and ten questions each.

  The seventh version of the afternoon examination is a general examination suitable for anyone, but in particular, for engineers whose specialties are not one of the other six disciplines. Though the subjects in the general afternoon examination correspond to the morning subjects, the questions are more complex—hence their double weighting. Questions on the afternoon examination are intended to cover concepts learned in the last two years of a four-year degree program. Unlike morning questions, these questions may deal with more than one basic concept per question.

  The percentages of questions for each subject in the general afternoon session examination are given in the following table.

Afternoon FE Exam Subjects (General Exam)

subject	percentage of questions (%)
chemistry	7.5
computers	5
dynamics	7.5
electrical circuits	10
engineering economics	5
ethics	5
fluid mechanics	7.5
materials science and structure of matter	5
mathematics	20
mechanics of materials	7.5
statics	10
thermodynamics	10

The percentages of questions for each subject in the civil discipline-specific afternoon session examination are as follows. The discipline-specific afternoon examinations cover substantially different bodies of knowledge than the morning examination. Formulas and tables of data needed to solve questions in these examinations will be included in either the NCEES Handbook or in the body of the question statement itself.

Afternoon FE Exam Subjects (DS Exam)

subject	percentage of questions (%)
computers and numerical methods	10
construction management	5
environmental engineering	10
hydraulics and hydrologic systems	10
legal and professional aspects	5
soil mechanics and foundations	10
structural analysis (frames, trusses, etc.)	10
structural design (concrete, steel, etc.)	10
surveying	10
transportation facilities	10
water purification and treatment	10

Each afternoon question consists of a problem statement followed by multiple-choice questions. Four answer choices lettered (A) through (D) are given, from which you must choose the best answer.

- Each question in the afternoon is worth two points, making the total possible score 120 points.
- The scores from the morning and afternoon sessions are added together to determine your total score. No points are subtracted for guessing or incorrect answers. Both sessions are given equal weight. It is not necessary to achieve any minimum score on either the morning or afternoon sessions.
- All grading is done by computer optical sensing.

Use of SI Units on the FE Exam

Metric questions are used in all subjects, except some civil engineering and surveying subjects that typically use only customary U.S. (i.e., English) units. SI units are consistent with ANSI/IEEE standard 268 (the American Standard for Metric Practice). Non-SI metric units might still be used when common or where needed for consistency with tabulated data (e.g., use of bars in pressure measurement).

Grading and Scoring the FE Exam

The FE exam is not graded on the curve, and there is no guarantee that a certain percent of examinees will pass. Rather, NCEES uses a modification of the Angoff procedure to determine the suggested passing score (the cutoff point or cut score).

With this method, a group of engineering professors and other experts estimate the fraction of minimally qualified engineers that will be able to answer each question correctly. The summation of the estimated fractions for all test questions becomes the passing score. The passing score in recent years has been somewhat less than 50 percent (i.e., a raw score of approximately 110 points out of 240). Because the law in most states requires engineers to achieve a score of 70 percent to become licensed, you may be reported as having achieved a score of 70 percent if your raw score is greater than the passing score established by NCEES, regardless of the raw percentage. The actual score may be slightly more or slightly less than 110 as determined from the performance of all examinees on the equating subtest.

Approximately 20 percent of each FE exam consists of questions repeated from previous examinations—this is the *equating subtest.* Since the performance of previous examinees on the equating subtest is known, comparisons can be made between the two examinations and examinee populations. These comparisons are used to adjust the passing score.

The individual states are free to adopt their own passing score, but all adopt NCEES's suggested passing score because the states believe this cutoff score can be defended if challenged.

You will receive the results approximately 12 to 14 weeks after the examination. If you pass, your score may or may not be revealed to you, depending on your state's policy, but if you fail, you will receive your score.

The following table lists the approximate fractions of examinees passing the FE exam.

Approximate FE Exam Passing Rates

category	percent passing
total, all U.S. states	60%–70%
ABET accredited, four-year engineering degrees[a]	70%–80%
nonaccredited, four-year engineering degrees	50%–65%
ABET accredited, four-year technology degrees[a]	35%–45%
nonaccredited, four-year technology degrees	25%–35%
nongraduates	35%–40%

[a]The Accreditation Board for Engineering and Technology (ABET) reviews and approves engineering degree programs in the United States. No engineering degree programs offered by universities outside of the United States and its territories or the Commonwealth of Puerto Rico are accredited by ABET.

Permitted Reference Material

Since October 1993, the FE examination has been what NCEES calls a "limited-reference" exam. This means that no books or references other than those supplied by NCEES may be used. Therefore, the FE examination is really an "NCEES-publication only" exam. NCEES provides its own FE Reference Handbook for use during the examination. No books from other publishers may be used.

CALCULATORS

In most states, battery- or solar-powered, silent calculators can be used, although printers cannot be used. (The solar-powered calculators are preferred because they do not have batteries that run down.) In most states, programmable, preprogrammed, or business/finance calculators are allowed. Similarly, nomographs and specialty slide rules are permitted. To prevent unauthorized transcription and redistribution of the examination questions, calculators with communication or text editing capabilities are banned from all NCEES exam sites. You cannot share calculators with other examinees.

It is essential that a calculator used for engineering examinations have the following functions.

- trigonometric functions
- inverse trigonometric functions
- hyperbolic functions
- pi
- square root and x^2
- common and natural logarithms
- y^x and e^x

For maximum speed, your calculator should also have or be programmed for the following functions.

- extracting roots of quadratic and higher-order equations
- converting between polar (phasor) and rectangular vectors
- finding standard deviations and variances
- calculating determinants of 3×3 matrices
- linear regression
- economic analysis and other financial functions

STRATEGIES FOR PASSING THE FE EXAM

The most successful strategy to pass the FE exam is to prepare in all of the examination subjects. Do not limit the number of subjects you study in hopes of finding enough questions in your particular areas of knowledge to pass.

Fast recall and stamina are essential to doing well. You must be able to quickly recall solution procedures, formulas, and important data. You will not have time during the exam to derive solutions methods—you must know them instinctively. This ability must be maintained for eight hours. Be sure to gain familiarity with the NCEES Handbook by using it as your only reference for some of the problems you work when you study.

In order to get exposure to all examination subjects, it is imperative that you develop and adhere to a review schedule. If you are not taking a classroom review course (where the order of your preparation is determined by the lectures), prepare your own review schedule.

There are also physical demands on your body during the examination. It is very difficult to remain alert and attentive for eight hours or more. Unfortunately, the more time you study, the less time you have to maintain your physical condition. Thus, most examinees arrive at the examination site in peak mental condition but in deteriorated physical condition. While preparing for the FE exam is not the only good reason for embarking on a physical conditioning program, it can serve as a good incentive to get in shape.

It will be helpful to make a few simple decisions prior to starting your review. You should be aware of the different options available to you. For example, you should decide early on to

- use SI units in your preparation
- perform electrical calculations with effective (rms) or maximum values
- take calculations out to a maximum of four significant digits
- prepare in all examination subjects, not just your specialty areas

At the beginning of your review program, you should locate a spare calculator. It is not necessary to buy a spare if you can arrange to borrow one from a friend or the office. However, if possible, your primary and spare calculators should be identical. If your spare calculator is not identical to the primary calculator, spend some time familiarizing yourself with its functions.

A Few Days Before the Exam

There are a few things you should do a week or so before the examination date. For example, visit the exam site in order to find the building, parking areas, examination room, and rest rooms. You should also make arrangements for child care and transportation. Since the examination does not always start or end at the designated times, make sure that your child care and transportation arrangements can tolerate a later-than-scheduled completion.

Next in importance to your scholastic preparation is the preparation of your two examination kits. The first kit consists of a bag or box containing items to bring with you into the examination room.

[ ] letter admitting you to the examination
[ ] photographic identification
[ ] main calculator
[ ] spare calculator
[ ] extra calculator batteries
[ ] a large eraser
[ ] unobtrusive snacks
[ ] travel pack of tissues
[ ] headache remedy
[ ] $2.00 in change
[ ] light, comfortable sweater
[ ] loose shoes or slippers
[ ] handkerchief
[ ] cushion for your chair
[ ] small hand towel
[ ] earplugs
[ ] wristwatch with alarm
[ ] wire coat hanger
[ ] extra set of car keys

The second kit consists of the following items and should be left in a separate bag or box in your car in case they are needed.

[ ] copy of your application
[ ] proof of delivery
[ ] this book
[ ] other references
[ ] regular dictionary
[ ] scientific dictionary
[ ] course notes in three-ring binders
[ ] cardboard box (use as a bookcase)
[ ] instruction booklets for all your calculators
[ ] light lunch
[ ] beverages in thermos and cans
[ ] sunglasses
[ ] extra pair of prescription glasses
[ ] raincoat, boots, gloves, hat, and umbrella
[ ] street map of the examination site
[ ] note to the parking patrol for your windshield explaining where you are, what you are doing, and why your time may have expired
[ ] battery-powered desk lamp

The Day Before the Exam

Take the day before the examination off from work to relax. Do not cram the last night. A good prior night's sleep is the best way to start the examination. If you live far from the examination site, consider getting a hotel room in which to spend the night.

Make sure your exam kits are packed and ready to go.

The Day of the Exam

You should arrive at least 30 minutes before the examination starts. This will allow time for finding a convenient parking place, bringing your materials to the examination room, and making room and seating changes. Be prepared, though, to find that the examination room is not open or ready at the designated time.

Once the examination has started, consider the following suggestions.

- Set your wristwatch alarm for five minutes before the end of each four-hour session and use that remaining time to guess at all of the remaining unsolved problems. Do not work up until the very end. You will be successful with about 25 percent of your guesses, and these points will more than make up for the few points you might earn by working during the last five minutes.

- Do not spend more than two minutes per morning question. (The average time available per problem is two minutes.) If you have not finished a question in that time, make a note of it and continue on.

- Do not ask your proctors technical questions. Even if they are knowledgeable in engineering, they will not be permitted to answer your questions.

- Make a quick mental note about any problems for which you cannot find a correct response or for which you believe there are two correct answers. Errors in the exam are rare, but they do occur. Being able to point out an error later might give you the margin you need to pass. Since such problems are almost always discovered during the scoring process and discounted from the examination, it is not necessary to tell your proctor, but be sure to mark the one best answer before moving on.

- Make sure all of your responses on the answer sheet are dark and completely fill the bubbles.

[ ] this book
[ ] other references
[ ] regular dictionary
[ ] scientific dictionary
[ ] course notes in three-ring binders
[ ] cardboard box (use as a bookcase)
[ ] instruction booklets for all your calculators
[ ] light lunch
[ ] beverages in thermos and cans
[ ] sunglasses
[ ] extra pair of prescription glasses
[ ] raincoat, boots, gloves, hat, and umbrella
[ ] street map of the examination site
[ ] note to the parking patrol for your windshield explaining where you are, what you are doing, and why your time may have expired
[ ] battery-powered desk lamp

The Day Before the Exam

Take the day before the examination off from work to relax. Do not cram the last night. A good prior night's sleep is the best way to start the examination. If you live far from the examination site, consider getting a hotel room in which to spend the night.

Make sure your exam kits are packed and ready to go.

The Day of the Exam

You should arrive at least 30 minutes before the examination starts. This will allow time for finding a convenient parking place, bringing your materials to the examination room, and making room and seating changes. Be prepared, though, to find that the examination room is not open or ready at the designated time.

Once the examination has started, consider the following suggestions.

- Set your wristwatch alarm for five minutes before the end of each four-hour session and use that remaining time to guess at all of the remaining unsolved problems. Do not work up until the very end. You will be successful with about 25 percent of your guesses, and these points will more than make up for the few points you might earn by working during the last five minutes.

- Do not spend more than two minutes per morning question. (The average time available per problem is two minutes.) If you have not finished a question in that time, make a note of it and continue on.

- Do not ask your proctors technical questions. Even if they are knowledgeable in engineering, they will not be permitted to answer your questions.

- Make a quick mental note about any problems for which you cannot find a correct response or for which you believe there are two correct answers. Errors in the exam are rare, but they do occur. Being able to point out an error later might give you the margin you need to pass. Since such problems are almost always discovered during the scoring process and discounted from the examination, it is not necessary to tell your proctor, but be sure to mark the one best answer before moving on.

- Make sure all of your responses on the answer sheet are dark and completely fill the bubbles.

Common Questions About the DS Exam

Q: Do I have to take the DS exam?

A: Most people do not have to take the DS exam and may elect the general exam option. The state boards do not care which afternoon option you choose; nor do employers. In some cases, examinees who are still in their undergraduate degree program may be required by their university to take a specific DS exam.

Q: Do all mechanical, civil, electrical, chemical, industrial, and environmental engineers take the DS exam?

A: Originally, the concept was that examinees from the "big five" disciplines would take the DS exam, and the general exam would be for everyone else. This remains just a concept, however. A majority of engineers in all of the disciplines apparently take the general exam.

Q: When do I elect to take the DS exam?

A: You will make your decision on the afternoon of the FE exam, when the exam booklet (containing all of the DS exams) is distributed to you.

Q: Where on the application for the FE exam do I choose which DS exam I want to take?

A: You don't specify the DS option at the time of your application.

Q: After starting to work on either the DS or general exam, can I change my mind and switch options?

A: Yes. Theoretically, if you haven't spent too much time on one exam, you can change your mind and start a different one. (You might need to obtain a new answer sheet from the proctor.)

Q: After I take the DS exam, does anyone know that I took it?

A: After you take the FE exam, only NCEES and your state board will know whether you took the DS or general exam. Such information may or may not be retained by your state board.

Q: Will my DS FE certificate be recognized by other states?

A: Yes. All states recognize passing the FE exam and do not distinguish between the DS and general afternoon portions of the FE exam.

Q: Is the DS FE certificate "better" than the general FE certificate?

A: There is no difference. No one will know which option you chose. It's not stated on the certificate you receive from your state.

Q: What is the format of the DS exam?

A: The DS exam is 4 hours long. There are 60 problems, each worth 2 points. The average time per problem is 4 minutes. Each problem is multiple choice with 4 answer choices. Most problems require the application of more than one concept (i.e., formula).

Q: Is there anything special about the way the DS exam is administered?

A: In all ways, the DS and general afternoon exam are equivalent. There is no penalty for guessing. No credit is given for scratch pad work, methods, etc.

Q: Are the answer choices close or tricky?

A: Answer choices are not particularly close together in value, so the number of significant digits is not going to be an issue. Wrong answers, referred to as "distractors" by NCEES, are credible. However, the exam is not "tricky"; it does not try to mislead you.

Q: Are any problems in the afternoon related to each other?

A: Several questions may refer to the same situation or figure. However, NCEES has tried to make all of the questions independent. If you make a mistake on one question, it shouldn't carry over to another.

Q: Is there any minimum passing score for the DS exam?

A: No. It is the total score from your morning and afternoon sessions that determines your passing, not the individual session scores. You do not have to "pass" each session individually.

Q: Is the general portion easier, harder, or the same as the DS exams?

A: Theoretically, all of the afternoon options are the same. At least, that is the intent of offering the specific options: to reduce the variability. Individual passing rates, however, may still vary 5 to 10 percent from exam to exam. (Professional Publications lists the most recent passing statistics for the various DS options on its website at www.ppi2pass.com.)

Q: Do the DS exams cover material at the undergraduate or graduate level?

A: Like the general exam, test topics come entirely from the typical undergraduate degree program. However, the emphasis is primarily on material from the third and fourth year of your program. This may put examinees who take the exam in their junior year at a disadvantage.

Q: Do you need practical work experience to take the DS exam?

A: No.

Q: Does the DS exam also draw on subjects that are in the general exam?

A: Yes. The dividing line between general and DS topics is often indistinct.

Q: Is the DS exam in customary U.S. or SI units?

A: The DS exam is essentially entirely in SI units. A few exceptions exist for some civil engineering subjects (surveying, hydrology, code-based design, etc.) where current common practice is limited to customary U.S. units.

Q: Does the NCEES Handbook cover everything that is on the DS exam?

A: No. You may be tested on subjects that are not present in the NCEES Handbook. However, NCEES has apparently adopted an unofficial policy of providing any necessary information, data, and formulas in the stem of the question. You will not be required to memorize any formulas.

Q: How is the DS reference material identified in the NCEES Handbook?

A: In most cases, the DS reference material is consolidated in the back. However, this policy is not consistently followed. In some cases, the DS reference material is mixed in with the general reference material. Only in some cases is this "intermixed" material identified as being DS material.

Q: Is everything in the DS portion of the NCEES Handbook going to be on the exam?

A: Apparently, there is a fair amount of reference material that isn't needed for every exam. There is no way, however, to know what material is needed.

Q: How long does it take to prepare for the DS exam?

A: Preparing for the DS exam is similar to preparing for a mini PE exam. Engineers typically take two to three months to complete a thorough review for the PE exam. However, examinees who are still in their degree program at a university probably aren't going to spend more than two weeks thinking about, worrying about, or preparing for the DS exam. They rely on their recent familiarity with the subject matter.

Q: If I take the DS exam and fail, do I have to take the DS exam the next time?

A: No. The examination process has no memory.

Q: Where can I get even more information about the DS exam?

A: If you have internet access, visit the Exam FAQs and the Engineering Exam Forum at Professional Publications' website. Our address is www.ppi2pass.com.

How to Use this Book

HOW EXAMINEES CAN USE THIS BOOK

This book is divided into two parts: The first part consists of 60 representative practice problems covering all of the topics in the afternoon DS exam. 60 problems happen to correspond to the number of problems in the afternoon DS exam. You may time yourself by allowing approximately 4 minutes per problem when attempting to solve these problems, but that was not my intent when designing this book. Since the solution follows directly after each problem in this section, I intended for you to read through the problems, attempt to solve them on your own, become familiar with the support material in the official NCEES Handbook, and accumulate the reference materials you think you will need for additional study.

The second part of this book consists of a complete sample examination that you can use as a source of additional practice problems or as a timed diagnostic tool. It also contains 60 problems, and the number of problems in each subject corresponds to the breakdown of subjects published by NCEES. Since the solutions to this part of the book are consolidated at the end, it was my intent that you would solve these problems in a realistic mock-exam mode.

You should use the NCEES Handbook as your only reference during this mock exam.

The morning general exam and the afternoon DS exam essentially cover two different bodies of knowledge. It takes a lot of discipline to prepare for two standardized exams simultaneously. Because of that (and because of my good understanding of human nature), I suspect that you will be tempted to start preparing for your chosen DS exam only after you have become comfortable with the general subjects. That's actually quite logical, because if you run out of time, you will still have the general afternoon exam as a viable option.

If, however, you are limited in time to only two or three months of study, it will be quite difficult to do a thorough DS review if you wait until after you have finished your general review. With a limited amount of time, you really need to prepare for both exams in parallel.

HOW INSTRUCTORS CAN USE THIS BOOK

The availability of the discipline-specific FE exam has greatly complicated the lives of review course instructors and coordinators. The general consensus is that it is essentially impossible to do justice to all of the general FE exam topics and then present a credible review for each of the DS topics. Increases in course cost, expenses, course length, and instructor pools (among many other issues) all conspire to create quite a difficult situation.

One-day reviews for each DS subject are subject-overload from a reviewing examinee's standpoint. Efforts to shuffle FE students over the parallel PE review courses meet with scheduling conflicts. Another idea, that of lengthening lectures and providing more in-depth coverage of existing topics (e.g., covering transistors during the electricity lecture), is perceived as a misuse of time by a majority of the review course attendees. Is it any wonder that virtually every FE review course in the country has elected to only present reviews for the general afternoon exam?

But, while more than half of the examinees elect to take the general afternoon exam, some may actually be required to take a DS exam. This is particularly the case in some university environments where the FE exam has become useful as an "outcome assessment tool." Thus, some method of review is still needed.

Since most examinees begin reviewing approximately two to three months before the exam (which corresponds to when most review courses begin), it is impractical to wait until the end of the general review to start the DS review. The DS review must proceed in parallel with the general review.

In the absence of parallel DS lectures (something that isn't yet occurring in too many review courses), you may want to structure your review course to provide lectures only on the general subjects. Your DS review could be assigned as "independent study," using chapters and problems from this book. Thus, your DS review would consist of distributing this book with a schedule of assignments. Your instructional staff could still provide assistance on specific DS problems, and completed DS assignments could still be recorded.

The final chapter on incorporating DS subjects into review courses has yet to be written. Like the landscape architect who waits until a well-worn path appears through the plants before placing stepping stones, we need to see how review courses do it before we can give any advice.

Practice Problems

COMPUTERS AND NUMERICAL METHODS

Problems 1 and 2 are based on the following information and illustration.

A porous sample is divided into a coarse mesh with nodes at the coordinates shown. Assume the heads, h, at either end of the sample are constant along the vertical faces.

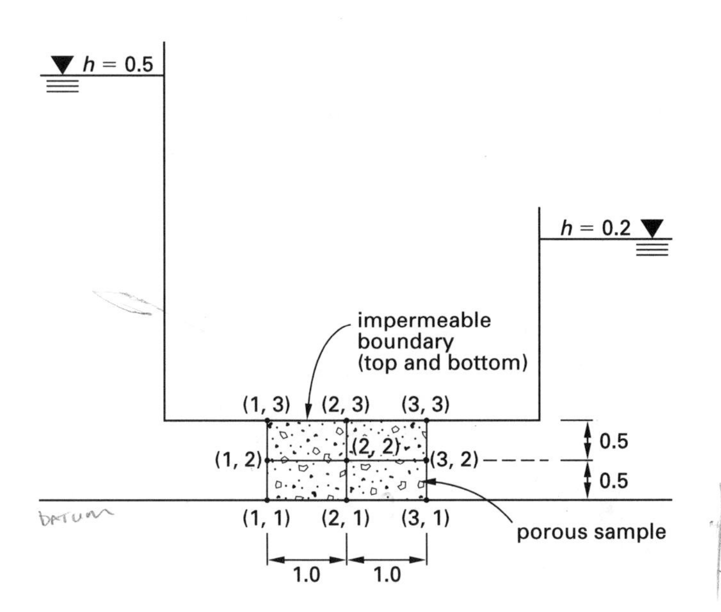

1. Using the finite difference method, the head at node (2, 2) is most nearly

(A) 0.25
(B) 0.35
(C) 0.40
(D) 0.45

h= 0.5 - 0.2

Solution:

Since the mesh is symmetrical, $h_{(2,1)} = h_{(2,3)}$. This leads to two equations and two unknowns, $h_{(2,2)}$ and $h_{(2,3)}$, which can be solved simultaneously.

$$h_{(2,2)} = \frac{h_{(1,2)} + h_{(3,2)} + h_{(2,1)} + h_{(2,3)}}{4} = \frac{h_{(1,2)} + h_{(3,2)} + 2h_{(2,3)}}{4}$$

$$4h_{(2,2)} - 2h_{(2,3)} = h_{(1,2)} + h_{(3,2)} = 0.5 + 0.2 = 0.7$$

$$4h_{(2,2)} - 2h_{(2,3)} = 0.7 \qquad \text{Eq. 1}$$

$$h_{(2,3)} = \frac{h_{(1,3)} + h_{(3,3)} + h_{(2,2)}}{3}$$

$$-h_{(2,2)} + 3h_{(2,3)} = h_{(1,3)} + h_{(3,3)} = 0.5 + 0.2 = 0.7$$

$$-h_{(2,2)} + 3h_{(2,3)} = 0.7 \qquad \text{Eq. 2}$$

Simultaneously solving Eqs. 1 and 2 gives $h_{(2,2)} = 0.35$ and $h_{(2,3)} = 0.35$.

Answer is B.

2. Using the finite difference method, the head at node (2, 3) is most nearly

(A) 0.25
(B) 0.35
(C) 0.40
(D) 0.45

Solution:

As determined in Prob. 1, $h_{(2,3)} = 0.35$.

Answer is B.

Problems 3 and 4 are based on the following information.

A civil engineer is debugging a computer program written in FORTRAN. The lines of a certain section of code are as follows.

```
C    INITIALIZE THE C ARRAY
     DO 3370 K = 1,5
        C(K) = 0
        D(K) = 0
3370 CONTINUE
     C(1) = 25.0
     D(1) = 1.0
     DO 3380 K = 2,5
        C(K) = C(K - 1)*D(K - 1) + D(K - 1)
        D(K) = D(K - 1) + C(K)
3380 CONTINUE
```

3. After executing this section of code, the value of C(5) is most nearly

(A) 5.5519×10^5
(B) 5.5264×10^5
(C) 3.05×10^{11}
(D) 3.50×10^{11}

Solution:

The initial values are

$$C(1) = 25.0$$
$$D(1) = 1.0$$

Subsequently,

K	C(K) = C(K − 1)*D(K − 1) + D(K − 1)	D(K) = D(K − 1) + C(K)
2	26.0	27.0
3	729.0	756.0
4	551,880.0	552,636.0
5	3.05×10^{11}	3.05×10^{11}

Answer is C.

4. After executing this section of code, the value of D(5) is most nearly

(A) 5.5519×10^5
(B) 5.5264×10^5
(C) 3.05×10^{11}
(D) 3.50×10^{11}

Solution:

As determined in Prob. 3, D(5) $= 3.05 \times 10^{11}$.

Answer is C.

Problems 5 and 6 are based on the following information.

The general equation for the central-difference approximation of the first derivative of a function, $y = f(x)$, with respect to x and evaluated at x_i, is given by

$$y_i' = \frac{y_{i+1} - y_{i-1}}{(2)(\Delta x)}$$

This equation is derived from the first three terms of the Taylor series expansions for function values at $x_{i+1} = x_i + \Delta x$ and $x_{i-1} = x_i - \Delta x$ as given by

$$y_{i+1} = y(x_i + \Delta x) = y_i + y_i'(\Delta x) + \frac{y_i''(\Delta x)^2}{2!}$$

$$y_{i-1} = y(x_i - \Delta x) = y_i - y_i'(\Delta x) + \frac{y_i''(\Delta x)^2}{2!}$$

5. Using the given Taylor series expansions, the general equation for the central-difference approximation for the second derivative, y'', is

(A) $y_i'' = \dfrac{y_{i+1} + 2y_i + y_{i-1}}{\Delta x}$

(B) $y_i'' = \dfrac{y_{i+1} - 2y_i + y_{i-1}}{(\Delta x)^2}$

(C) $y_i'' = \dfrac{y_{i+1} + 2y_i + y_{i-1}}{(\Delta x)^2}$

(D) $y_i'' = \dfrac{y_{i+1} - y_i + y_{i-1}}{(\Delta x)^2}$

Solution:

The first derivative approximation was derived by subtracting the bottom Taylor series expression from the top Taylor series expression.

Similarly, to find the second derivative, the two Taylor series expressions are added together to give

$$y_{i+1} + y_{i-1} = 2y_i + (2)\left[\frac{y_i''(\Delta x)^2}{2!}\right]$$

Rearrangement of this expression gives

$$y_i'' = \frac{y_{i+1} - 2y_i + y_{i-1}}{(\Delta x)^2}$$

Answer is B.

6. If a function is given by $y = f(x) = e^x/(e+x)^x$ and an interval of $\Delta x = 0.5$ is used, then the value of the second derivative, y'', at $x = 2$ is most nearly

(A) 0.05
(B) 0.10
(C) 0.20
(D) 1.5

Solution:

For an interval of $\Delta x = 0.5$, a given location of $x_i = 2$, and a given function of $y = f(x) = e^x/(e+x)^x$,

$$x_{i+1} = x_i + \Delta x = 2 + 0.5 = 2.5$$

$$x_{i-1} = x_i - \Delta x = 2 - 0.5 = 1.5$$

$$y_i = f(2) = \frac{e^2}{(e+2)^2} = 0.3319$$

$$y_{i+1} = f(2.5) = \frac{e^{2.5}}{(e+2.5)^{2.5}} = 0.1958$$

$$y_{i-1} = f(1.5) = \frac{e^{1.5}}{(e+1.5)^{1.5}} = 0.5173$$

The general equation for the central-difference approximation as determined in Prob. 5 can be used to give

$$y_i'' = \frac{y_{i+1} - 2y_i + y_{i-1}}{(\Delta x)^2}$$

$$= \frac{0.1958 - (2)(0.3319) + 0.5173}{(0.5)^2}$$

$$= 0.1972 \quad (0.20)$$

Answer is C.

CONSTRUCTION MANAGEMENT

7. What is the following type of model called?

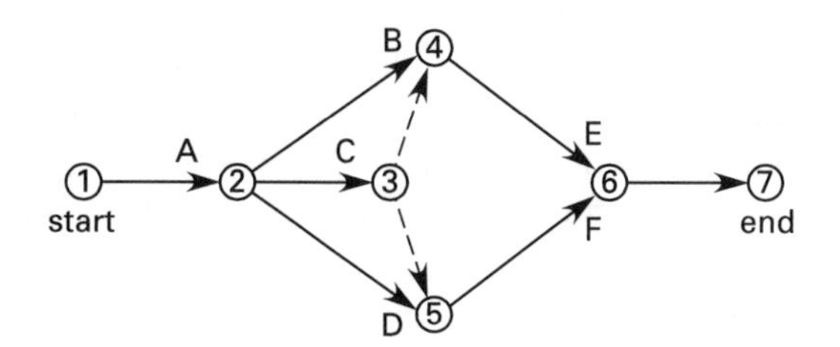

(A) a bubble (activity-on-node) network
(B) an arrow (activity-on-arrow) network
(C) a PERT chart
(D) a bar (Gantt) chart

Solution:

This is called an arrow (activity-on-arrow) network.

Answer is B.

8. What is the following type of model called?

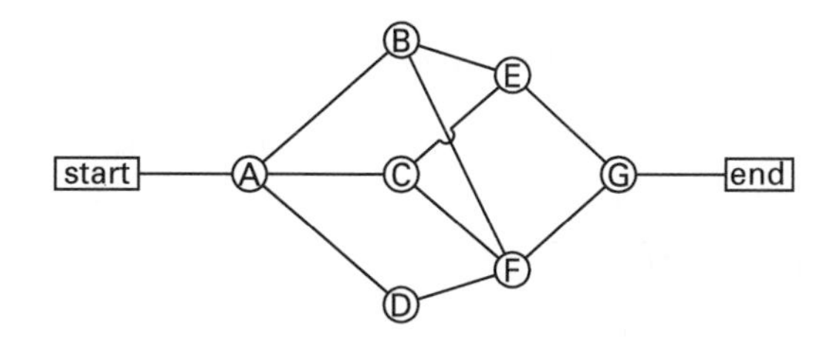

(A) a bubble (activity-on-node) network
(B) an arrow (activity-on-arrow) network
(C) a PERT chart
(D) a bar (Gantt) chart

Solution:

This is called a "bubble" (activity-on-node) network.

Answer is A.

9. A construction project is composed of activities A through H with the durations, in days, given for each activity in the diagram shown. This project is on a strict schedule that must be maintained and is scheduled to start at the beginning of January 1. Work can only be performed during the day and must be done on every day of the week (Sunday through Saturday).

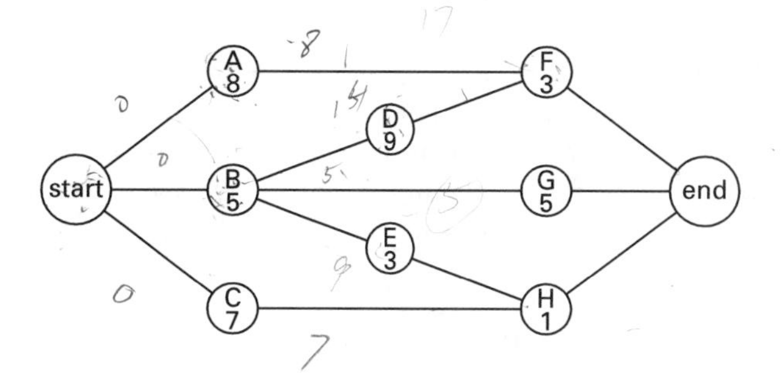

The critical path for this project passes through

(A) START-A-F-END
(B) START-C-H-END
(C) START-B-D-F-END
(D) START-B-E-H-END

Solution:

Problem 9 can be solved with critical path method (CPM) calculations. Since the project is to start on January 1, it is easy to designate the actual start time as the end of the previous day, December 31, and designate it as day 0 with January 1 being designated as day 1.

Determination of the earliest start time (EST) and earliest finish time (EFT) for an activity is done by a forward pass through the diagram.

The EST of an activity is calculated as the maximum of the EFTs of the activities preceding it. For example, activity A has no activities preceding it, so it has an EST of 0. The EFT of this activity is given by

$$\text{EFT of activity A} = \text{EST} + \text{duration} = 0 + 8 = 8$$

Similarly, activity B has an EST of 0 and an EFT of 5.

Activity D is preceded only by activity B, so its EST is the EFT of activity B. That is, the EST of activity D is 5 and the EFT is 14.

Activity F is preceded by two activities, A and D, so the EST of activity F is the maximum EFT of the two. That is, the EST of activity F is 14 and its EFT is 17.

Similar calculations show that the minimum project duration is 17 days (the EFT for activity F is 17).

Determination of the latest start time (LST) and latest finish time (LFT) for an activity is done by a backward pass through the diagram using the maximum project duration as the starting point. That is, 17 is the starting point for these calculations.

The LFT of an activity is the minimum LST of the activities following it. For example, the LFTs of the activities preceding the project finish are all 17 since no activities follow them. Accordingly, the LFTs of activities F and G are each 17. The LST of activity F is given by

$$\text{LST of activity F} = \text{LFT} - \text{duration} = 17 - 3 = 14$$

Similarly, the LST of activity G is $17 - 5 = 12$.

Activity D is preceded only by activity F, so its LFT is the LST of activity F. That is, the LFT of activity D is 14 and the LST of activity D is 5.

Activity B is preceded by two activities, D and G, so its LFT is the minimum LST of the two. That is, the LFT of activity B is 5.

The total float time (TF) of an activity is determined by either subtracting the EFT from the LFT or subtracting the EST from the LST. For example, the TF of activity G is

$$\text{TF of activity G} = \text{LFT} - \text{EFT} = 17 - 10 = 7$$

Making a summary table helps to organize these results.

activity	duration	EST	EFT	LST	LFT	TF
A	8	0	8	6	14	6
B	5	0	5	0	5	0
C	7	0	7	9	16	9
D	9	5	14	5	14	0
E	3	5	8	13	16	8
F	3	14	17	14	17	0
G	5	5	10	12	17	7
H	1	8	9	16	17	8

The critical path is the path that passes through the activities that result in a TF of 0. From the summary table, this passes through nodes B-D-F.

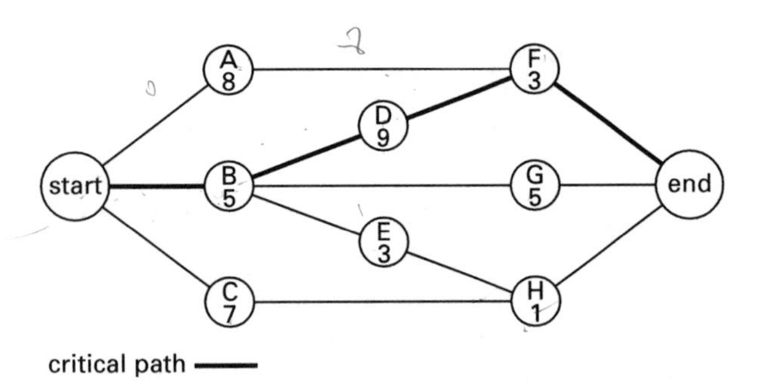

Answer is C.

ENVIRONMENTAL ENGINEERING

10. One liter of a solution is made by adding 3 g of acetic acid (HAc) to distilled water. The molecular weight of acetic acid is 60.052. The acid-dissociation constant for acetic acid is $K_A = 1.75 \times 10^{-5}$. The chemical equation for the dissociation of acetic acid into hydrogen ions and acetate ions is $\text{HAc} \longleftrightarrow \text{H}^+ + \text{Ac}^-$. The percentage of acetic acid ionized in solution is most nearly

(A) 0.4%
(B) 1.9%
(C) 5.0%
(D) 9.3%

Solution:

The molar concentration of the acid solution is

$$[\text{HAc}] = \frac{\dfrac{3\ \text{g}}{60.052\ \dfrac{\text{g}}{\text{mol}}}}{1\ \text{L}} = 0.05\ \text{mol/L} \quad (0.05\ \text{M})$$

By letting x equal the number of moles of acetic acid that ionize in solution to form hydrogen ions (H^+) and acetate ions (Ac^-), the appropriate relationship can be determined to calculate the amount of acetate ions in solution.

$$[\text{HAc}] = 0.05\ \frac{\text{mol}}{\text{L}} - x$$

$$[\text{H}^+] = [\text{Ac}^-] = x \quad [\text{in mol/L}]$$

$$\frac{[\text{H}^+][\text{Ac}^-]}{[\text{HAc}]} = K_A$$

$$\frac{(x)(x)}{0.05 - x} = 1.75 \times 10^{-5}$$

Solving for x gives the concentration of hydrogen or acetate ions in solution.

$$[\text{H}^+] = [\text{Ac}^-] = x = 9.27\times10^{-4}\ \text{mol/L} \quad (9.27\times10^{-4}\ \text{M})$$

The percentage of acetic acid ionized in solution is

$$\left(\frac{9.27 \times 10^{-4}\ \text{M}}{0.05\ \text{M}}\right)(100\%) = 1.854\% \quad (1.9\%)$$

Answer is B.

11. The chloride (Cl^-) concentration in a lake is found to be 10^{-2} M. The $\text{HgCl}_2(\text{aq})$ concentration is found to be 10^{-7} M. The following chemical equations and equilibrium constants apply.

$$\text{Hg}^{2+} + \text{Cl}^- \longleftrightarrow \text{HgCl}^+ \quad K_1 = 5.6 \times 10^6$$

$$\text{HgCl}^+ + \text{Cl}^- \longleftrightarrow \text{HgCl}_2 \quad K_2 = 3.0 \times 10^6$$

The concentration of Hg^{2+} is most nearly

(A) 5.9×10^{-17} M
(B) 3.3×10^{-12} M
(C) 3.0×10^{6} M
(D) 5.6×10^{6} M

Solution:

From the given chemical data,

$$\frac{[HgCl_2]}{[HgCl^+][Cl^-]} = K_2 = 3.0 \times 10^6$$

$$[HgCl^+] = \frac{[HgCl_2]}{(K_2)[Cl^-]} = \frac{10^{-7}}{(3.0 \times 10^6)(10^{-2})} = 3.33 \times 10^{-12} \text{ M}$$

$$\frac{[HgCl^+]}{[Hg^{2+}][Cl^-]} = K_1 = 5.6 \times 10^6$$

$$[Hg^{2+}] = \frac{[HgCl^+]}{(K_1)[Cl^-]} = \frac{3.33 \times 10^{-12}}{(5.6 \times 10^6)(10^{-2})} = 5.9 \times 10^{-17} \text{ M}$$

Answer is A.

12. Nickel is removed from a water stream with a pH of 9 by hydroxide precipitation. The atomic weight of nickel is 58.70. The chemical equation and solubility-product constant for this reaction are

$$Ni^{2+} + 2OH^- \longrightarrow Ni(OH)_2(s) \quad K_{sp} = 5.54 \times 10^{-16}$$

The solubility of Ni^{2+} in this water is most nearly

(A) 0.006 mg/L
(B) 0.33 mg/L
(C) 0.55 mg/L
(D) 0.59 mg/L

Solution:

From the given pH of 9, the OH^- concentration can be determined.

$$\begin{aligned} pH + pOH &= 14 \\ pOH &= 14 - pH \\ -\log[OH^-] &= 14 - pH = 14 - 9 = 5 \\ [OH^-] &= 1 \times 10^{-5} \text{ mol/L} \end{aligned}$$

From the chemical equation and solubility-product constant,

$$[Ni^{2+}][OH^-]^2 = K_{sp} = 5.54 \times 10^{-16}$$

$$[Ni^{2+}] = \frac{K_{sp}}{[OH^-]^2} = \frac{5.54 \times 10^{-16}}{(1 \times 10^{-5})^2} = 5.54 \times 10^{-6} \text{ mol/L}$$

Therefore, the solubility of nickel is

$$\left(5.54 \times 10^{-6} \frac{\text{mol}}{\text{L}}\right)\left(58.70 \frac{\text{g}}{\text{mol}}\right)\left(\frac{1000 \text{ mg}}{1 \text{ g}}\right) = 0.325 \text{ mg/L} \quad (0.33 \text{ mg/L})$$

Answer is B.

13. The 10,000 gal aeration basin shown maintains a constant 1500 mg/L mixed liquor suspended solids (MLSS) and treats 25,000 gal of liquid waste per day. The suspended solids are separated in a clarifier with recycle of separated sludge. The recycle flow rate is 5000 gal per day. Each day, 500 gal of recycle are wasted. The effluent from the clarifier contains a constant 30 mg/L MLSS. Assume steady-state flow conditions.

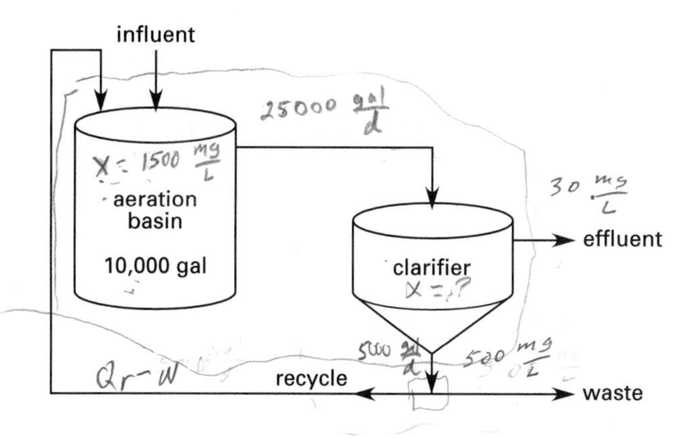

The solids residence time is most nearly

(A) 2 days
(B) 3 days
(C) 4 days
(D) 5 days

Solution:

The variables for flow rates, Q, and MLSS concentrations, X, can be placed on the schematic to help organize the solution to this problem.

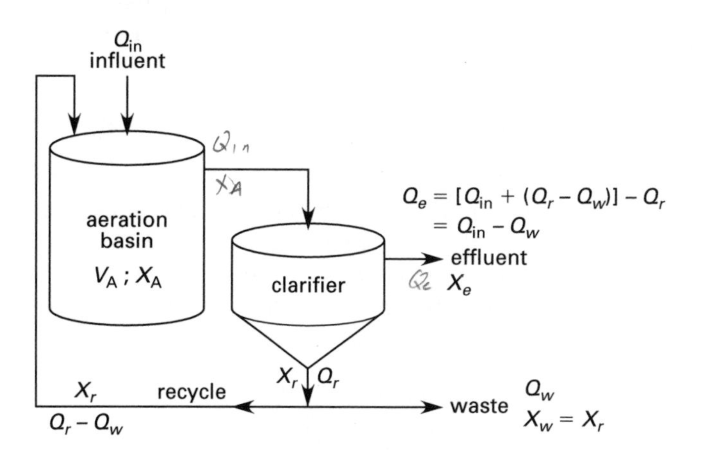

In the labeled schematic,

$$\begin{aligned} V_A &= 10{,}000 \text{ gal} \\ X_A &= 1500 \text{ mg/L} \\ X_e &= 30 \text{ mg/L} \\ Q_{\text{in}} &= 25{,}000 \text{ gal/day} \\ Q_w &= 500 \text{ gal/day} \\ Q_r &= 5000 \text{ gal/day} \\ X_w &= X_r \quad \text{[unknown]} \end{aligned}$$

To determine X_r, a solids balance must be taken at the clarifier. Since the total solids entering the clarifier must be equal to the total solids exiting the clarifier for steady-state flow conditions,

$$\begin{aligned} [Q_{\text{in}} + (Q_r - Q_w)]X_A &= Q_e X_e + Q_r X_r \\ &= (Q_{\text{in}} - Q_w)X_e + Q_r X_r \end{aligned}$$

$$X_r = \frac{[Q_{\text{in}} + (Q_r - Q_w)]X_A - (Q_{\text{in}} - Q_w)X_e}{Q_r}$$

$$= \frac{\left[25{,}000 \ \frac{\text{gal}}{\text{day}} + \left(5000 \ \frac{\text{gal}}{\text{day}} - 500 \ \frac{\text{gal}}{\text{day}}\right)\right]\left(1500 \ \frac{\text{mg}}{\text{L}}\right) - \left(25{,}000 \ \frac{\text{gal}}{\text{day}} - 500 \ \frac{\text{gal}}{\text{day}}\right)\left(30 \ \frac{\text{mg}}{\text{L}}\right)}{5000 \ \frac{\text{gal}}{\text{day}}}$$

$$= 8703 \text{ mg/L}$$

By noting that $X_w = X_r$, the solids residence time can be calculated to be

$$\frac{V_A X_A}{Q_w X_w + Q_e X_e} = \frac{V_A X_A}{Q_w X_r + (Q_{\text{in}} - Q_w)X_e}$$

$$= \frac{(10{,}000 \text{ gal})\left(1500 \ \frac{\text{mg}}{\text{L}}\right)}{\left(500 \ \frac{\text{gal}}{\text{day}}\right)\left(8703 \ \frac{\text{mg}}{\text{L}}\right) + \left(25{,}000 \ \frac{\text{gal}}{\text{day}} - 500 \ \frac{\text{gal}}{\text{day}}\right)\left(30 \ \frac{\text{mg}}{\text{L}}\right)}$$

$$= 2.95 \text{ days} \quad (3.0 \text{ days})$$

Answer is B.

14. A water sample from a stream with an average flow of 95 000 L per day contains 225 mg/L of cyanide waste in the form of sodium cyanide (NaCN). Chlorine can be added to the stream to destroy this NaCN waste according to the reaction

$$\begin{aligned} &2\text{NaCN} + 5\text{Cl}_2 + 12\text{NaOH} \longrightarrow \\ &\qquad \text{N}_2 + \text{Na}_2\text{CO}_3 + 10\text{NaCl} + 6\text{H}_2\text{O} \end{aligned}$$

All atomic weights are found from a periodic table of the chemical elements to be Na =22.98977, C =12.011, N =14.0067, Cl = 35.453, O = 15.9994, and H = 1.0079.

The theoretical minimum amount of chlorine required to destroy the NaCN waste is most nearly

(A) 80 kg/day
(B) 160 kg/day
(C) 170 kg/day
(D) 200 kg/day

Solution:

The 225 mg/L concentration of cyanide can be expressed as 225 parts per million (ppm) = 225 parts/10^6 parts.

The total mass of NaCN flowing past a given stream cross section per day is

$$\dot{m} = \dot{V}\rho \times \text{concentration}$$

$$= \frac{\left(95{,}000 \ \frac{\text{L}}{\text{day}}\right)\left(1000 \ \frac{\text{kg}}{\text{m}^3}\right)\left(\frac{225 \text{ kg}}{1 \times 10^6 \text{ kg}}\right)}{1000 \ \frac{\text{L}}{\text{m}^3}}$$

$$= 21.4 \text{ kg/day}$$

Relevant molecular weights are

$$\begin{aligned} \text{MW}_{\text{NaCN}} &= 22.98977 \ \frac{\text{g}}{\text{mol}} + 12.011 \ \frac{\text{g}}{\text{mol}} + 14.0067 \ \frac{\text{g}}{\text{mol}} \\ &= 49.007 \text{ g/mol} \end{aligned}$$

$$\begin{aligned} \text{MW}_{\text{Cl}_2} &= (2)\left(35.453 \ \frac{\text{g}}{\text{mol}}\right) \\ &= 70.906 \text{ g/mol} \end{aligned}$$

The number of moles of NaCN flowing past a given stream cross section each day is

$$\frac{21.4 \ \frac{\text{kg}}{\text{day}}}{\left(49.007 \ \frac{\text{g}}{\text{mol}}\right)\left(\frac{1 \text{ kg}}{1000 \text{ g}}\right)} = 437 \text{ mol/day NaCN}$$

From the given chemical equation, the destruction of 2 mol of NaCN requires 5 mol of chlorine (Cl_2). Therefore, the amount of Cl_2 required to destroy the given amount of NaCN each day is

$$\left(\frac{5 \text{ mol Cl}_2}{2 \text{ mol NaCN}}\right)\left(437 \ \frac{\text{mol NaCN}}{\text{day}}\right) \times \left(70.906 \ \frac{\text{g}}{\text{mol}}\right)\left(\frac{1 \text{ kg}}{1000 \text{ g}}\right) = 77.5 \text{ kg Cl}_2\text{/day} \quad (80 \text{ kg/day})$$

Answer is A.

15. A proposed landfill is to be 400 m by 200 m in plan area and 25 m deep. The average daily filling rate is expected to be 15 m by 10 m by 3 m deep, and the daily cover to be used is 0.2 m thick. Assume that the landfill will be operational every week from Monday through Friday.

The projected life of the landfill is most nearly

(A) 16 years
(B) 17 years
(C) 20 years
(D) 23 years

Solution:

The total volume of the proposed landfill is

$$V_{\text{total}} = (400 \text{ m})(200 \text{ m})(25 \text{ m}) = 2 \times 10^6 \text{ m}^3$$

The rate of trash input into the landfill is

$$\dot{V}_{\text{trash}} = \frac{(15 \text{ m})(10 \text{ m})(3 \text{ m})}{1 \text{ day}} = 450 \text{ m}^3/\text{day}$$

The rate of fill from use of the daily cover is

$$\dot{V}_{\text{cover}} = \frac{(15 \text{ m})(10 \text{ m})(0.2 \text{ m})}{1 \text{ day}} = 30 \text{ m}^3/\text{day}$$

Since the landfill is to be operational for 5 days each week and there are 52 weeks in the year, the projected life of the landfill is

$$\begin{aligned} t_{\text{landfill}} &= \frac{V_{\text{total}}}{\dot{V}_{\text{trash}} + \dot{V}_{\text{cover}}} \\ &= \left(\frac{2 \times 10^6 \text{ m}^3}{450 \, \frac{\text{m}^3}{\text{day}} + 30 \, \frac{\text{m}^3}{\text{day}}} \right) \left(\frac{1 \text{ wk}}{5 \text{ days}} \right) \left(\frac{1 \text{ yr}}{52 \text{ wk}} \right) \\ &= 16.03 \text{ years} \quad (16 \text{ years}) \end{aligned}$$

Answer is A.

HYDRAULICS AND HYDROLOGIC SYSTEMS

16. A manometer is shown with $h_p = 25$ cm and $h_m = 63$ cm. The pipe fluid is oil with a specific gravity of 0.8. Mercury has a specific gravity of 13.6. Assume standard conditions.

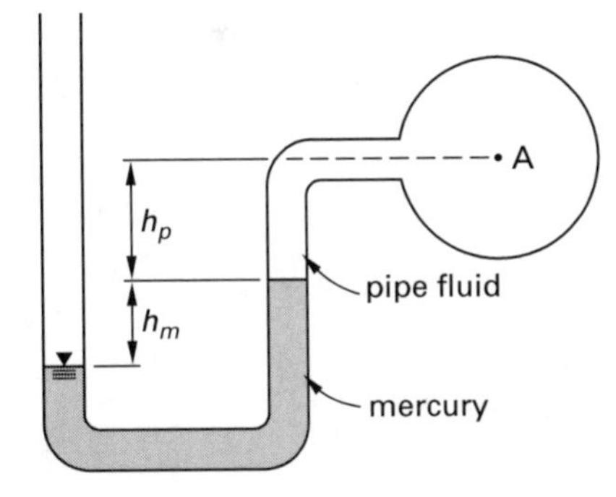

The gage pressure at point A is most nearly

(A) −80 kPa
(B) −86 kPa
(C) −88 kPa
(D) −90 kPa

Solution:

The gage pressure at point B is zero. The mass density of water is $\rho_w = 1000 \text{ kg/m}^3$ at standard conditions.

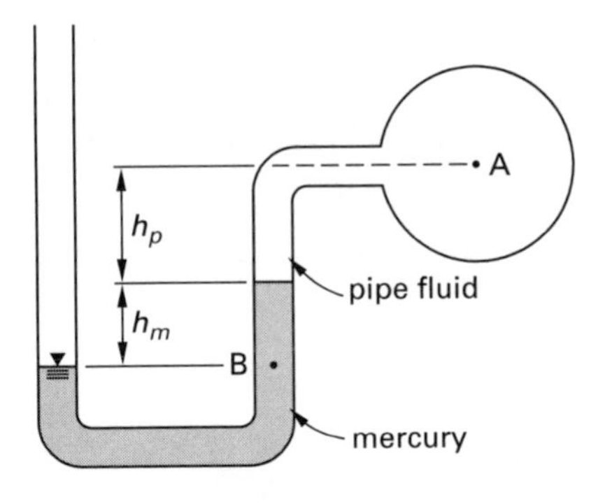

Therefore, from equilibrium,

$$\begin{aligned} p_{\text{B}} = 0 &= \gamma_{\text{oil}} h_p + \gamma_{\text{Hg}} h_m + p_{\text{A}} \\ &= (\text{SG})_{\text{oil}} \rho_w g h_p + (\text{SG})_{\text{Hg}} \rho_w g h_m + p_{\text{A}} \end{aligned}$$

This can be rearranged to solve for gage pressure at point A.

$$\begin{aligned} p_{\text{A}} &= -[(\text{SG})_{\text{oil}} \rho_w g h_p + (\text{SG})_{\text{Hg}} \rho_w g h_m] \\ &= -\left[\begin{array}{l} (0.8) \left(1000 \, \frac{\text{kg}}{\text{m}^3}\right) \left(9.81 \, \frac{\text{m}}{\text{s}^2}\right) \left(\dfrac{25 \text{ cm}}{100 \, \frac{\text{cm}}{\text{m}}} \right) \\ + (13.6) \left(1000 \, \frac{\text{kg}}{\text{m}^3}\right) \left(9.81 \, \frac{\text{m}}{\text{s}^2}\right) \left(\dfrac{63 \text{ cm}}{100 \, \frac{\text{cm}}{\text{m}}} \right) \end{array} \right] \\ &= -86\,014 \text{ Pa} \quad (86 \text{ kPa}) \end{aligned}$$

Answer is B.

17. The Rational Formula runoff coefficient of a 300 m long by 200 m wide property with a 3% slope is 0.35. The rainfall intensity is 116 mm/h.

The discharge from this property is most nearly

(A) 2200 m^3/h
(B) 2400 m^3/h
(C) 3800 m^3/h
(D) 7000 m^3/h

Solution:

The discharge from this property is

$$\begin{aligned} Q &= ciA \\ &= (0.35)\left(116\ \frac{\text{mm}}{\text{h}}\right)\left(\frac{1\ \text{m}}{1000\ \text{mm}}\right)(300\ \text{m})(200\ \text{m}) \\ &= 2436\ \text{m}^3/\text{h} \quad (2400\ \text{m}^3/\text{h}) \end{aligned}$$

Answer is B.

18. A concrete sanitary sewer is 150 m long and has a pipe diameter of 1.25 m. The inlet elevation is 50.0 m, and the outlet elevation is 49.0 m. The Manning roughness coefficient, assumed to be constant with depth of flow, is 0.012. During heavy rainfalls, the sewer pipe flows full with no surcharge.

During heavy rainfalls, the capacity of the sewer is most nearly

(A) 3.1 m^3/s
(B) 3.8 m^3/s
(C) 4.7 m^3/s
(D) 5.7 m^3/s

Solution:

The slope is calculated to be

$$\begin{aligned} S &= \frac{\text{inlet elev} - \text{outlet elev}}{\text{pipe length}} = \frac{50.0\ \text{m} - 49.0\ \text{m}}{150\ \text{m}} \\ &= 0.00667 \end{aligned}$$

Since the pipe flows full during heavy rainfalls, the wetted perimeter is the entire perimeter of the pipe. The hydraulic radius is calculated to be

$$R = \frac{\text{area}}{\text{wetted perimeter}} = \frac{\frac{\pi(1.25\ \text{m})^2}{4}}{\pi(1.25\ \text{m})} = 0.3125\ \text{m}$$

From Manning's equation, the velocity of flow is

$$\begin{aligned} \text{v} &= \left(\frac{1}{n}\right)R^{\frac{2}{3}}S^{\frac{1}{2}} = \left(\frac{1}{0.012}\right)(0.3125\ \text{m})^{\frac{2}{3}}(0.00667)^{\frac{1}{2}} \\ &= 3.13\ \text{m/s} \end{aligned}$$

The flow capacity is then

$$\begin{aligned} Q = \text{v}A &= \left(3.13\ \frac{\text{m}}{\text{s}}\right)\left(\frac{\pi(1.25\ \text{m})^2}{4}\right) \\ &= 3.84\ \text{m}^3/\text{s} \quad (3.8\ \text{m}^3/\text{s}) \end{aligned}$$

Answer is B.

Problems 19 and 20 are based on the following information.

Water is pumped from a lake with a pipe inlet at an elevation of 200 m to a tank at an elevation of 205 m. The pipeline from the lake to the tank is 300 m long and is cast iron, with a 30 cm inside pipe diameter. The pump efficiency is 80%. Minor losses, entrances losses, and exit losses are negligible. The flow rate through the piping is 1.25 m^3/s. Assume steady, incompressible flow. The kinematic viscosity of water is $1 \times 10^{-6}\ m^2/s$. The roughness factor for cast iron is $e = 0.25$ mm.

19. Using the Darcy equation, the head loss in the piping is most nearly

(A) 300 m
(B) 310 m
(C) 320 m
(D) 330 m

Solution:

The roughness factor for cast iron is $e = 0.25$ mm. The relative roughness is

$$\begin{aligned} \text{relative roughness} &= \frac{e}{D} = \frac{0.25\ \text{mm}}{(30\ \text{cm})\left(\frac{10\ \text{mm}}{1\ \text{cm}}\right)} \\ &= 0.000833 \end{aligned}$$

The area of flow is

$$A = \frac{\pi D^2}{4} = \frac{\pi\left((30\ \text{cm})\left(\frac{1\ \text{m}}{100\ \text{cm}}\right)\right)^2}{4} = 0.07069\ \text{m}^2$$

The Reynolds number is

$$\begin{aligned} \text{Re} &= \frac{\text{v}D}{\nu} = \frac{\left(\frac{Q}{A}\right)D}{\nu} \\ &= \frac{\left(\frac{1.25\ \frac{\text{m}^3}{\text{s}}}{0.07069\ \text{m}^2}\right)(30\ \text{cm})\left(\frac{1\ \text{m}}{100\ \text{cm}}\right)}{1 \times 10^{-6}\ \frac{\text{m}^2}{\text{s}}} \\ &= 5.305 \times 10^6 \end{aligned}$$

From the Moody diagram for the calculated e/D and Re, the friction factor is $f \approx 0.0188$.

Therefore, from the Darcy equation, the head loss in the piping is

$$\begin{aligned} h_f &= f\left(\frac{L}{D}\right)\left(\frac{\text{v}^2}{2g}\right) = f\left(\frac{L}{D}\right)\left[\frac{\left(\frac{Q}{A}\right)^2}{2g}\right] \\ &= (0.0188)\left[\frac{300\ \text{m}}{(30\ \text{cm})\left(\frac{1\ \text{m}}{100\ \text{cm}}\right)}\right] \\ &\quad \times \left[\frac{\left(\frac{1.25\ \frac{\text{m}^3}{\text{s}}}{0.07069\ \text{m}^2}\right)^2}{(2)\left(9.81\ \frac{\text{m}}{\text{s}^2}\right)}\right] \\ &= 299.6\ \text{m}\quad (300\ \text{m}) \end{aligned}$$

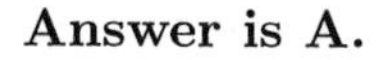

Answer is A.

20. The power provided by the pump to raise water from the lake to the tank at the flow rate indicated is most nearly

(A) 3.0 MW
(B) 3.8 MW
(C) 4.7 MW
(D) 5.4 MW

Solution:

The total head required to lift the fluid is

$$\begin{aligned} h &= (\text{tank elev} - \text{lake elev}) + \text{pipe head loss} \\ &= (205\ \text{m} - 200\ \text{m}) + 300\ \text{m} \\ &= 305\ \text{m} \end{aligned}$$

The input power required by the pump to provide the required head is

$$\begin{aligned} \dot{W} &= \frac{Q\gamma_w h}{\eta} = \frac{Q\rho_w g h}{\eta} \\ &= \frac{\left(1.25\ \frac{\text{m}^3}{\text{s}}\right)\left(1000\ \frac{\text{kg}}{\text{m}^3}\right)\left(9.81\ \frac{\text{m}}{\text{s}^2}\right)(305\ \text{m})}{0.80} \\ &= 4.675\times 10^6\ \text{W}\quad (4.7\ \text{MW}) \end{aligned}$$

Answer is C.

21. A reservoir with a water surface level at an elevation of 200 m drains through a 1 m diameter pipe with the outlet at an elevation of 180 m. The pipe outlet discharges to atmospheric pressure. The total head losses in the pipe and fittings are 18 m. Assume steady, incompressible flow.

The flow rate out of the pipe outlet is most nearly

(A) 4.9 m^3/s
(B) 6.3 m^3/s
(C) 31 m^3/s
(D) 39 m^3/s

Solution:

Using the pipe outlet as the datum, the variables in the energy equation are as follows.

$$\begin{aligned} p_1 &= 0 \quad \left[\begin{array}{c}\text{reservoir free surface is}\\ \text{at atmospheric pressure}\end{array}\right] \\ z_1 &= 200\ \text{m} \\ \text{v}_1 &\approx 0 \quad \left[\begin{array}{c}\text{water has negligible}\\ \text{velocity at reservoir surface}\end{array}\right] \\ p_2 &= 0 \quad \left[\begin{array}{c}\text{pipe outlet discharges}\\ \text{to atmospheric pressure}\end{array}\right] \\ z_2 &= 180\ \text{m} \end{aligned}$$

The energy equation can be rearranged to solve for the velocity at the pipe outlet.

$$\frac{p_1}{\gamma} + z_1 + \frac{\text{v}_1^2}{2g} = \frac{p_2}{\gamma} + z_2 + \frac{\text{v}_2^2}{2g} + h_{f\text{total}}$$

$$\begin{aligned} \text{v}_2 &= \sqrt{2g\left(\frac{p_1}{\gamma} + z_1 + \frac{\text{v}_1^2}{2g} - \frac{p_2}{\gamma} - z_2 - h_{f\text{total}}\right)} \\ &= \sqrt{2g\left(\frac{p_1}{\rho g} + z_1 + \frac{\text{v}_1^2}{2g} - \frac{p_2}{\rho g} - z_2 - h_{f\text{total}}\right)} \\ &= \sqrt{(2)\left(9.81\ \frac{\text{m}}{\text{s}^2}\right)\left[\begin{array}{l}0 + 200\ \text{m} + 0\\ \quad - 0 - 180\ \text{m} - 18\ \text{m}\end{array}\right]} \\ &= 6.26\ \text{m/s} \end{aligned}$$

The flow rate out of the pipe outlet is

$$\begin{aligned} Q = \text{v}A &= \left(6.26\ \frac{\text{m}}{\text{s}}\right)\left(\frac{\pi(1\ \text{m})^2}{4}\right) \\ &= 4.92\ \text{m}^3/\text{s}\quad (4.9\ \text{m}^3/\text{s}) \end{aligned}$$

Answer is A.

LEGAL AND PROFESSIONAL ASPECTS

22. An engineering company is retained for services. Shortly into the project, the company's chief engineer, who is a registered professional engineer, realizes that a potential conflict of interest exists between her responsibilities to the company and her responsibilities to the client.

Which of the following ethical standards apply?

(A) The chief engineer's primary responsibilities are to the company she works for. Therefore, she should continue working on the project with no change in plans.
(B) The chief engineer should inform her employer and the company's client of any business association, interest, or other circumstances that could influence her professional judgment or the quality of her work.
(C) The chief engineer should write a letter to the registration board notifying them of her situation.
(D) Both B and C apply.

Solution:

Answer is B.

23. John Smith, an independent registered professional engineer, is checking the work for two different clients on the same project. His first client is designing the substructure of a bridge project, and his second is designing the superstructure of the same bridge project.

Under which of the following provisions can Smith be compensated by his clients for his work?

(A) He can only be paid by the client designing the bridge substructure since this is the part of the structure that is first constructed and, usually, first paid for.
(B) He can only be paid by the client designing the bridge superstructure since this is the main component of the bridge.
(C) He can be paid by both clients as long as circumstances are fully disclosed to, and agreed upon, by all interested parties.
(D) He cannot accept compensation directly from either client and must notify the registration board about his situation so that the board can collect the compensation for his work and, in turn, submit it to him.

Solution:

Answer is C.

24. The term "conceptual issues" in an engineering code of ethics, according to NCEES, refers to

(A) issues that are raised among registration board members as possible concepts for charting the future of the engineering profession
(B) issues that are raised among registered professional engineers as possible concepts for charting the future of the engineering profession
(C) circumstances in which terminology in the professional engineering code of ethics is not clear
(D) pre-design concepts discussed with a client to determine optimal engineering solutions

Solution:

Answer is C.

SOILS AND FOUNDATIONS

25. A constant head permeameter is shown. The soil is homogeneous, isotropic, and saturated.

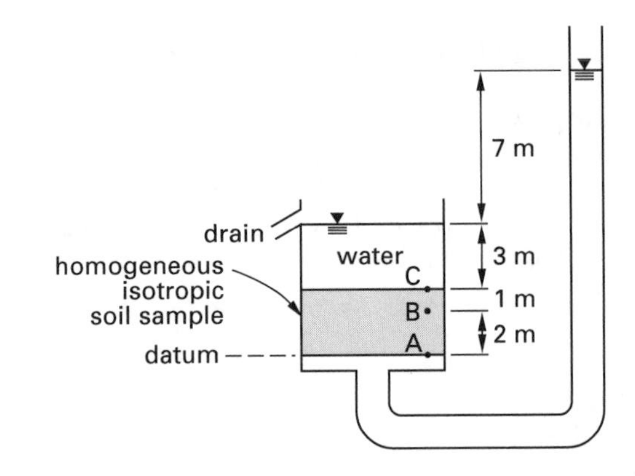

The pressure head at point B in the soil sample is most nearly

(A) 6.00 m
(B) 6.33 m
(C) 6.67 m
(D) 6.75 m

Solution:

The total head, h_t, is the sum total of the elevation and pressure heads as given by $h_t = h_e + h_p$. This can be rearranged to give $h_p = h_t - h_e$.

A tabulation of known heads with respect to the datum is as follows.

point	h_e (m)	h_p (m)	h_t (m)
A	0	13	13
B	2	?	?
C	3	3	6

Since the soil sample is saturated, homogeneous, and isotropic, and since the steady-state continuity equation requires constant flow velocity through the soil, the head gradient through the soil is linear.

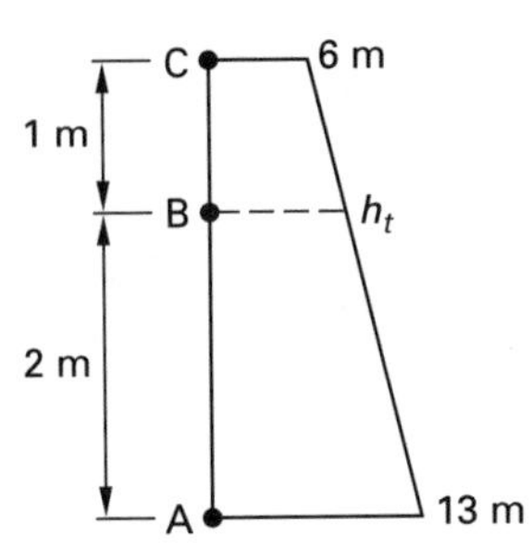

At point B, the total head can be found by similar triangles.

$$\frac{h_t - 6\text{ m}}{1\text{ m}} = \frac{13\text{ m} - 6\text{ m}}{3\text{ m}}$$

$$h_t = \left(\frac{13\text{ m} - 6\text{ m}}{3\text{ m}}\right)(1\text{ m}) + 6\text{ m}$$
$$= 8.33\text{ m}$$

The pressure head, h_p, at point B is given by

$$h_p = h_t - h_e = 8.33\text{ m} - 2\text{ m} = 6.33\text{ m}$$

Answer is B.

Problems 26 and 27 are based on the following information and illustration.

A retaining wall extends from the top of bedrock to the ground surface. A resisting force, F, is located on the opposite side of the wall to provide support. A frictionless hinge at point A prevents the base of the wall from sliding. Soil is homogeneous, isotropic, and cohesionless.

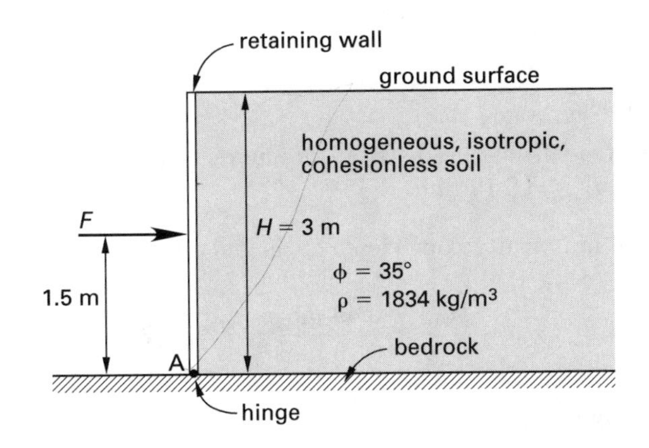

26. Using Rankine's theory, the total active resultant lateral earth force per unit length of retaining wall is most nearly

(A) 15 kN/m
(B) 22 kN/m
(C) 44 kN/m
(D) 82 kN/m

Solution:

From Rankine's theory, the coefficient of active lateral earth pressure for cohesionless soils is given by

$$K_a = \tan^2\left(45° - \frac{\phi}{2}\right) = \tan^2\left(45° - \frac{35°}{2}\right) = 0.27$$

The active lateral earth pressure distribution is linear.

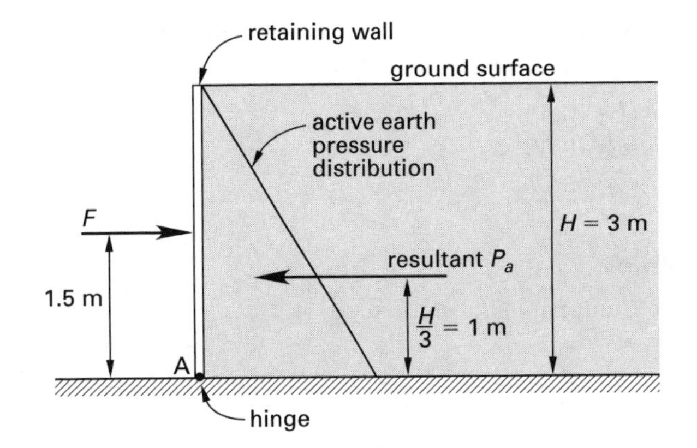

The active lateral earth pressure at any depth, h, below the ground surface can be found by

$$\sigma_a = K_a\sigma_v = K_a\gamma h = K_a\rho g h$$

From this, the total force is the resultant, P_a, as determined by finding the total area under the active earth pressure profile.

$$P_a = \tfrac{1}{2}K_a\gamma H^2 = \tfrac{1}{2}K_a\rho g H^2$$
$$= \left(\frac{1}{2}\right)(0.27)\left(1834\ \frac{\text{kg}}{\text{m}^3}\right)\left(9.81\ \frac{\text{m}}{\text{s}^2}\right)(3\text{ m})^2$$
$$= 21\,860\text{ N/m}\quad(22\text{ kN/m})$$

Answer is B.

27. Assuming that wall friction is negligible, the minimum required force, F, per unit length of retaining wall to resist the overturning moment is most nearly

(A) 15 kN/m
(B) 22 kN/m
(C) 44 kN/m
(D) 82 kN/m

Solution:

The resultant, P_a, acts at a distance of $H/3 = 1$ m from the base of the wall. From Prob. 26, $P_a = 22$ kN per meter of wall length. Summing moments about point A gives

$$F = \frac{P_a\left(\frac{H}{3}\right)}{1.5 \text{ m}} = \frac{\left(22 \ \frac{\text{kN}}{\text{m}}\right)(1 \text{ m})}{1.5 \text{ m}} = 14.67 \text{ kN/m}$$

Answer is A.

28. A soil sample has a total mass of 23.3 g, a volume of 12 cm^3, an oven-dry mass of 21.2 g, and a specific gravity of 2.5 for the solids.

The void ratio of this soil sample is most nearly

(A) 0.42
(B) 0.53
(C) 0.62
(D) 0.71

Solution:

For this problem, soil is modeled as a three-phase system.

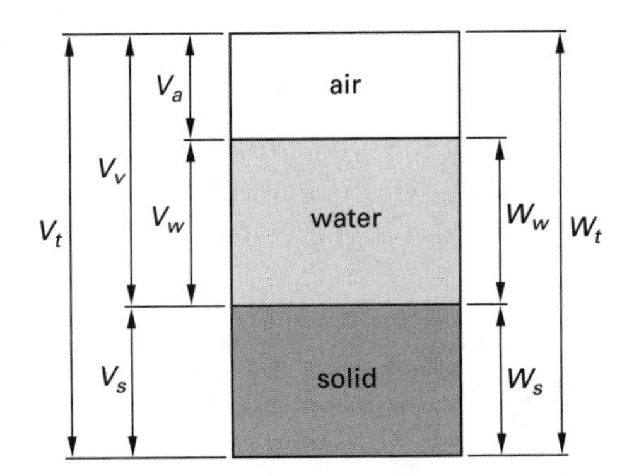

In this model,

V_t = total volume
V_v = volume of voids
V_a = volume of air
V_w = volume of water
V_s = volume of solids
W_t = total weight or mass
W_w = weight or mass of water
W_s = weight or mass of solids

The volume of soil solids, V_s, is given by

$$V_s = \frac{W_s}{G\rho_w} = \frac{21.2 \text{ g}}{(2.5)\left(1 \ \frac{\text{g}}{\text{cm}^3}\right)} = 8.48 \text{ cm}^3$$

$[\rho_w = \text{density of water}]$

The volume of the voids is found by subtracting the volume of solids from the total volume.

$$V_v = V_t - V_s = 12 \text{ cm}^3 - 8.48 \text{ cm}^3 = 3.52 \text{ cm}^3$$

The void ratio, e, is given by

$$e = \frac{V_v}{V_s} = \frac{3.52 \text{ cm}^3}{8.48 \text{ cm}^3} = 0.415 \quad (0.42)$$

Answer is A.

Problems 29 and 30 are based on the following information and illustration.

A soil's grain-size distribution curve is as shown.

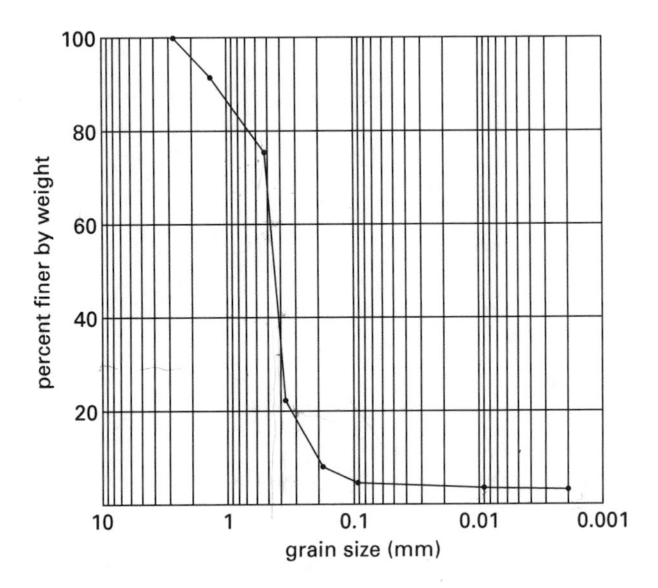

29. The uniformity coefficient is most nearly

(A) 1.6
(B) 2.1
(C) 2.6
(D) 3.2

Solution:

As read from the distribution curve, $D_{60} = 0.49$ mm, and $D_{10} = 0.19$ mm.

The uniformity coefficient, c_u, is given by

$$c_u = \frac{D_{60}}{D_{10}} = \frac{0.49 \text{ mm}}{0.19 \text{ mm}} = 2.58 \quad (2.6)$$

Answer is C.

30. The coefficient of gradation is most nearly

(A) 0.17
(B) 0.44
(C) 1.6
(D) 3.0

Solution:

As read from the distribution curve, $D_{30} = 0.39$ mm. The values of D_{60} and D_{10} are from Prob. 29.

The coefficient of gradation, c_c, is given by

$$c_c = \frac{(D_{30})^2}{(D_{60})(D_{10})} = \frac{(0.39 \text{ mm})^2}{(0.49 \text{ mm})(0.19 \text{ mm})} = 1.63 \quad (1.6)$$

Answer is C.

STRUCTURAL ANALYSIS

Problems 31–34 are based on the following information and illustration.

A plane truss is loaded as shown.

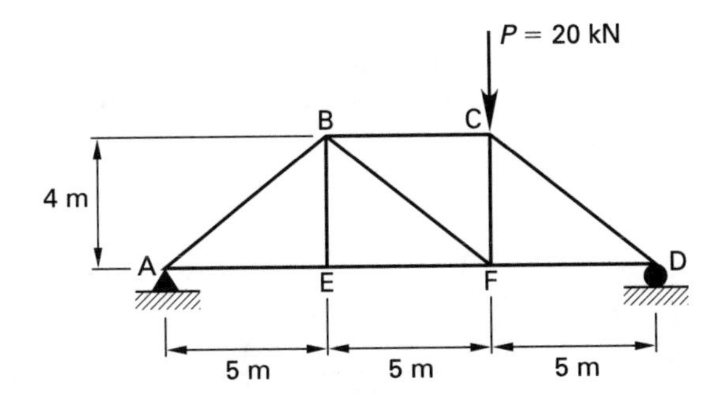

31. The magnitude of the vertical reaction force at support A is most nearly

(A) 3.3 kN
(B) 6.7 kN
(C) 10 kN
(D) 16 kN

Solution:

Since support D is a roller support, the horizontal reaction force, R_{Ax}, is zero. To find the vertical reaction at support A, R_{Ay}, a free-body diagram is drawn of the entire truss and moments are summed about support D.

$$R_{Ay}(15 \text{ m}) - (20 \text{ kN})(5 \text{ m}) = 0$$

$$R_{Ay} = \frac{(20 \text{ kN})(5 \text{ m})}{15 \text{ m}} = 6.67 \text{ kN} \quad (6.7 \text{ kN})$$

Answer is B.

32. The magnitude of the force in member AB is most nearly

(A) 4.2 kN
(B) 6.7 kN
(C) 8.5 kN
(D) 11 kN

Solution:

From Prob. 31, the reaction at support A is 6.67 kN. Next, a free-body diagram is drawn for support A with member forces and their force components. Using the Pythagorean theorem, the relative magnitudes of each force and each force's horizontal and vertical components can be found. (In this case, 6.403, 5, and 4 are the relative magnitudes of the member AB force and the forces's horizontal and vertical components, respectively.)

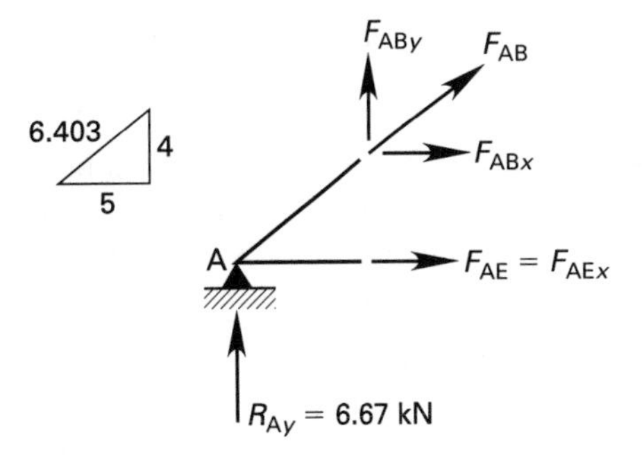

For equilibrium, all forces on a free-body must sum to zero. Therefore, summation of vertical forces gives $F_{ABy} + R_{Ay} = 0$, which can be rearranged to give $F_{ABy} = -R_{Ay} = -6.667$ kN. Recall that joints in trusses are frictionless so no bending moments exist.

The force and its components are proportional to the geometric lengths of the triangle sides.

$$F_{AB} = \left(\frac{6.403 \text{ m}}{4 \text{ m}}\right) F_{ABy} = \left(\frac{6.403 \text{ m}}{4 \text{ m}}\right)(-6.667 \text{ kN})$$
$$= -10.68 \text{ kN} \quad (-11 \text{ kN})$$

(Note that the answer is negative. This means that the calculated force is in the opposite direction to the assumed force direction on the free-body diagram. Since the member AB force was assumed to apply tension on the free-body diagram, the negative answer means that member AB is in compression.)

Answer is D.

33. The magnitude of the force in member EF is most nearly

(A) 4.2 kN
(B) 5.3 kN
(C) 6.7 kN
(D) 8.3 kN

Solution:

From Prob. 32, $F_{\text{AB}y} = -6.67$ kN. The horizontal component of the member AB force is

$$F_{\text{AB}x} = \left(\frac{5\text{ m}}{4\text{ m}}\right) F_{\text{AB}y} = \left(\frac{5\text{ m}}{4\text{ m}}\right)(-6.667\text{ kN})$$
$$= -8.33\text{ kN}$$

For equilibrium at support A, summation of the horizontal forces must be equal to zero. Therefore, the force in member AE is

$$F_{\text{AE}} = F_{\text{AE}x} = -F_{\text{AB}x} = -(-8.33\text{ kN}) = 8.33\text{ kN}$$

The positive sign in the calculated member AE force means that the assumed direction of the force on the free-body diagram, which indicates tension, is in the same direction as the calculated force.

A free-body diagram of joint E will show that the vertical member BE is unable to sustain any horizontal force. Therefore, the force in member EF is the same as the force in member AE.

$$F_{\text{EF}} = F_{\text{AE}} = 8.33\text{ kN}\quad(8.3\text{ kN})$$

Answer is D.

34. If the truss members are made of steel and the cross-sectional area of each member is 1000 mm^2, the magnitude of the vertical deflection at joint E is most nearly

(A) 0.7 mm
(B) 1.6 mm
(C) 2.3 mm
(D) 2.8 mm

Solution:

Use the principle of virtual work.

The actual forces in each member can be determined by applying the equations of equilibrium to each truss joint and are shown as follows. (Note that some roundoff error exists in these calculated numbers.)

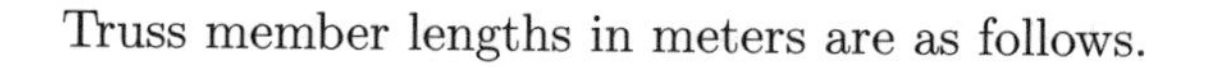

Truss member lengths in meters are as follows.

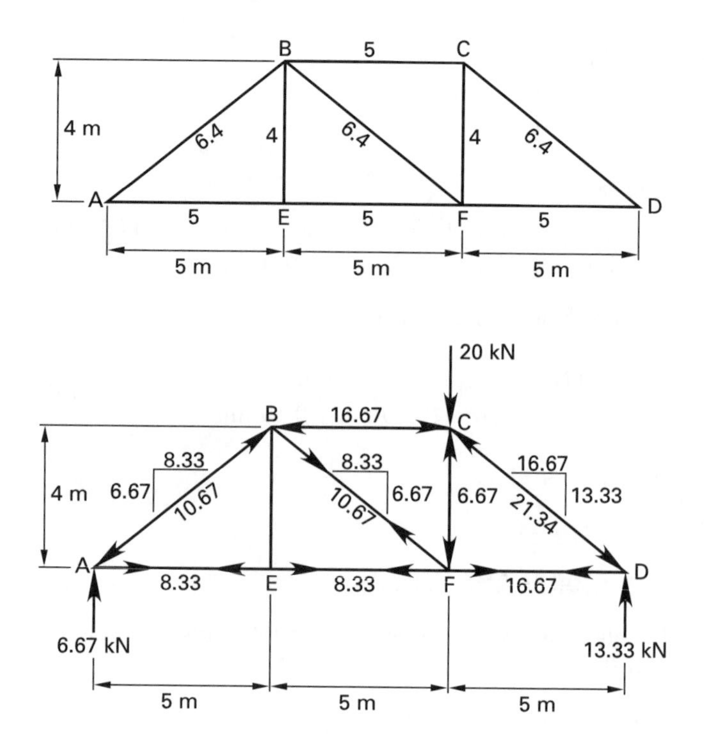

Application of a vertical unit load at joint E results in the virtual member forces as follows. (Note that some roundoff error exists in these calculated numbers.)

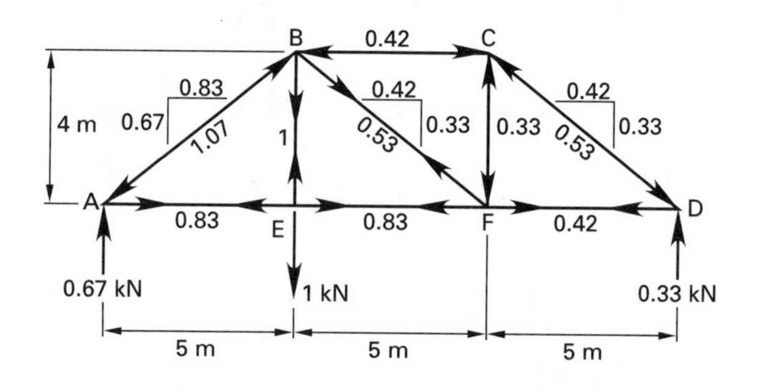

It is recommended that a table be set up to keep all variables organized.

member	F_Q (kN)	F_P (kN)	L (m)	F_QF_PL (kN^2·m)
AB	10.67	1.07	6.4	73.1
BC	16.67	0.42	5.0	35.0
CD	21.34	0.53	6.4	72.4
AE	8.33	0.83	5.0	34.6
EF	8.33	0.83	5.0	34.6
FD	16.67	0.42	5.0	35.0
BE	0.00	1.00	4.0	0.0
CF	6.67	0.33	4.0	8.8
BF	10.67	0.53	6.4	36.2
				$\sum = 329.7$

The modulus of elasticity of steel is $E = 2.1 \times 10^{11}$ Pa. Since the area and modulus of elasticity are the same for all truss members, AE is common to all members and can be taken outside of the summation for simplification. Therefore, the vertical deflection at point E can be found by

$$\Delta_E = \sum F_Q \delta L = \sum F_Q \left(\frac{F_P L}{AE}\right)$$
$$= \left(\frac{1}{AE}\right) \sum F_Q F_P L$$
$$= \left(\frac{329.7\ \text{kN}^2\cdot\text{m}}{(1000\ \text{mm}^2)\left(\frac{1\ \text{m}}{1000\ \text{mm}}\right)^2 \times (2.1 \times 10^{11}\ \text{Pa})\left(\frac{1\ \text{kN}}{1000\ \text{N}}\right)}\right)\left(\frac{1000\ \text{mm}}{1\ \text{m}}\right)$$
$$= 1.57\ \text{mm} \quad (1.6\ \text{mm})$$

Since the unit load in the virtual force system was downward and the answer is positive in sign, the actual deflection is also downward.

Answer is B.

Problems 35 and 36 are based on the following information and illustration.

A beam is loaded as shown.

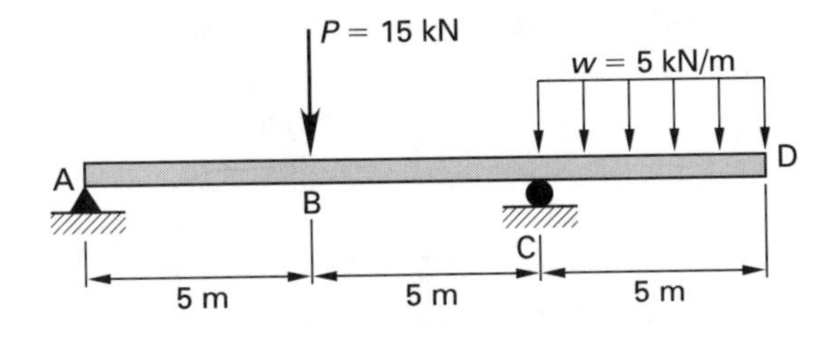

35. The magnitude of the vertical reaction force at support A is most nearly

(A) 1.3 kN
(B) 5.0 kN
(C) 13 kN
(D) 20 kN

Solution:

Since support C is a roller support, there is no horizontal reaction force at that point. The vertical reaction force at support A, R_{Ay}, can be found by converting the uniformly distributed load, w, into a resultant point load.

$$W = wL = \left(5\ \frac{\text{kN}}{\text{m}}\right)(5\ \text{m}) = 25\ \text{kN}$$

This resultant load is located at the centroid of the uniformly distributed load.

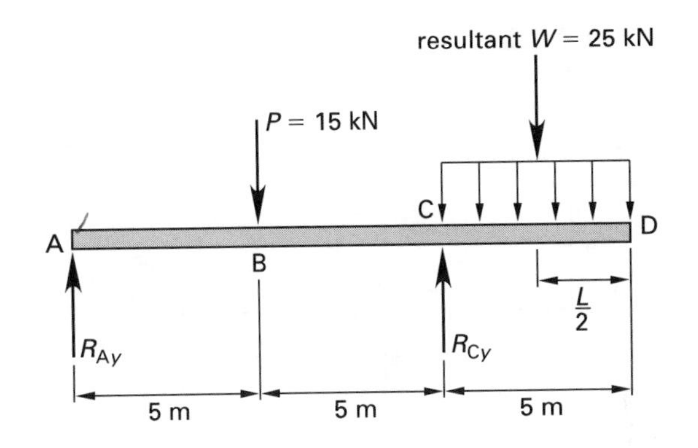

The support A vertical reaction can then be found by summing moments about support C. This results in a vertical reaction at support A of

$$R_{Ay}(10\ \text{m}) - (15\ \text{kN})(5\ \text{m}) + (25\ \text{kN})(2.5\ \text{m}) = 0$$

$$R_{Ay} = \frac{(15\ \text{kN})(5\ \text{m}) - (25\ \text{kN})(2.5\ \text{m})}{10\ \text{m}}$$
$$= 1.25\ \text{kN} \quad (1.3\ \text{kN})$$

Since the answer is positive in sign, the direction of the calculated reaction is the same as that of the assumed reaction; that is, the direction of the reaction is upward.

Answer is A.

36. The magnitude of the maximum vertical shear in the beam is most nearly

(A) 1.3 kN
(B) 14 kN
(C) 25 kN
(D) 39 kN

Solution:

One way to determine the answer to this problem is to construct a shear diagram. The change in shear is the area under the applied loading. Although a moment diagram is not required for this problem, it follows that the change in moment is the area under the shear diagram, so a moment diagram is usually included.

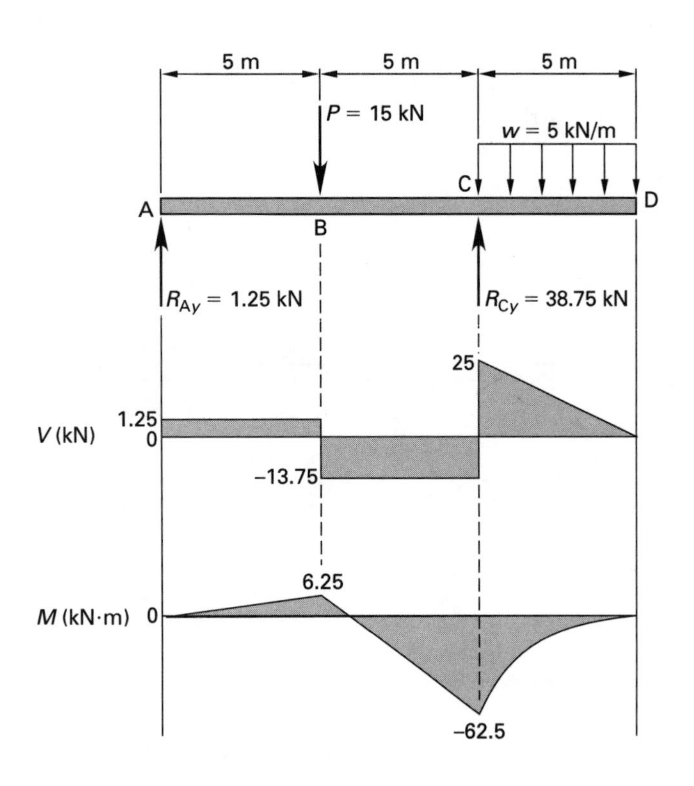

As can be seen, the maximum value of vertical shear is 25 kN at support C.

Answer is C.

STRUCTURAL DESIGN

Problems 37 and 38 are based on the following information and illustration.

The cross section of a reinforced concrete beam with tension reinforcement is shown. Assume that the beam is underreinforced.

$$f'_c = 3000 \text{ lbf/in}^2$$
$$f_y = 40{,}000 \text{ lbf/in}^2$$
$$A_s = 3 \text{ in}^2$$

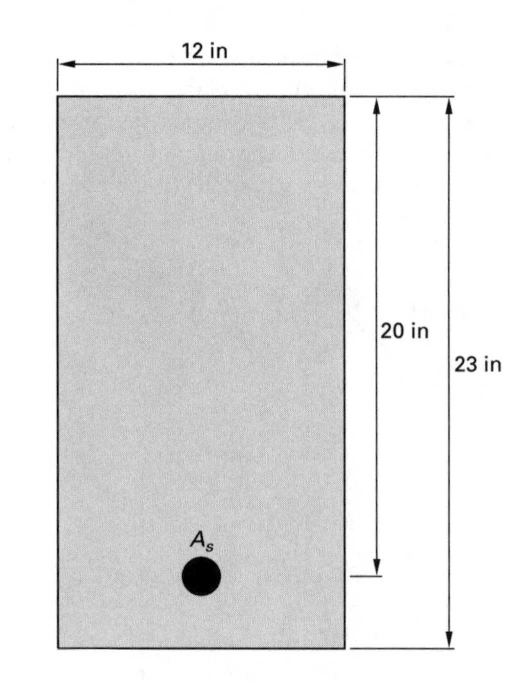

37. In accordance with American Concrete Institute (ACI) strength design, the allowable moment capacity of the beam is most nearly

(A) 162 ft-kips
(B) 181 ft-kips
(C) 201 ft-kips
(D) 213 ft-kips

Solution:

$$\rho = \frac{A_s}{bd} = \frac{3 \text{ in}^2}{(12 \text{ in})(20 \text{ in})} = 0.0125$$

In an actual design and analysis situation, a check should *always* be made to see that the actual reinforcing steel ratio falls between the allowable maximum and allowable minimum steel ratios, even though this check is not required to solve this specific problem. $\beta = 0.85$ since $f'_c \leq 4000 \text{ lbf/in}^2$.

The balanced and maximum allowable steel ratios are given by

$$\rho_b = \beta\left(\frac{0.85 f'_c}{f_y}\right)\left(\frac{87{,}000}{87{,}000 + f_y}\right)$$
$$= (0.85)\left[\frac{(0.85)\left(3000 \ \frac{\text{lbf}}{\text{in}^2}\right)}{40{,}000 \ \frac{\text{lbf}}{\text{in}^2}}\right] \times \left(\frac{87{,}000}{87{,}000 + 40{,}000 \ \frac{\text{lbf}}{\text{in}^2}}\right)$$
$$= 0.0371$$

$$\rho_{\text{max}} = 0.75\rho_b = (0.75)(0.0371) = 0.0278$$

The minimum allowable steel ratio is

$$\rho_{\min} = \frac{200}{f_y} = \frac{200}{40{,}000\ \frac{\text{lbf}}{\text{in}^2}} = 0.005$$

$$0.005 \leq 0.0125 \leq 0.0278$$

$$\rho_{\min} \leq \rho \leq \rho_{\max}$$

Therefore, the actual reinforcing steel ratio falls within allowable limits and conforms to ACI specifications.

The depth of the concrete compressive stress block is given by

$$a = \frac{A_s f_y}{0.85 f'_c b} = \frac{(3\ \text{in}^2)\left(40{,}000\ \frac{\text{lbf}}{\text{in}^2}\right)}{(0.85)\left(3000\ \frac{\text{lbf}}{\text{in}^2}\right)(12\ \text{in})} = 3.92\ \text{in}$$

For flexure, $\phi = 0.90$. The allowable moment capacity is

$$\begin{aligned} M_{\text{allow}} &= \phi M_n = \phi\left[0.85 f'_c a b\left(d - \frac{a}{2}\right)\right] \\ &= (0.90)\left[\begin{array}{l}(0.85)\left(3000\ \frac{\text{lbf}}{\text{in}^2}\right)(3.92\ \text{in}) \\ \times (12\ \text{in})\left(20\ \text{in} - \frac{3.92\ \text{in}}{2}\right)\end{array}\right] \\ &\quad \times \left(\frac{1\ \text{kip}}{1000\ \text{lbf}}\right)\left(\frac{1\ \text{ft}}{12\ \text{in}}\right) \\ &= 162.3\ \text{ft-kips} \end{aligned}$$

Answer is A.

38. If the dead load shear force in the beam, V_{dead}, is 5 kips and the live load shear force in the beam, V_{live}, is 15 kips, then the minimum amount of shear reinforcement needed for a center-to-center stirrup spacing of 12 in based on ACI strength design is most nearly

(A) 0.0010 in^2
(B) 0.0012 in^2
(C) 0.18 in^2
(D) 0.30 in^2

Solution:

For shear, $\phi = 0.85$ as specified by the American Concrete Institute (ACI).

The ultimate shear force in the beam is

$$\begin{aligned} V_u &= 1.4 V_{\text{dead}} + 1.7 V_{\text{live}} \\ &= (1.4)(5\ \text{kips}) + (1.7)(15\ \text{kips}) \\ &= 32.5\ \text{kips} \end{aligned}$$

The nominal concrete shear strength is

$$\begin{aligned} V_c &= 2\sqrt{f'_c}\, bd \\ &= (2)\left(\sqrt{3000\ \frac{\text{lbf}}{\text{in}^2}}\right)(12\ \text{in})(20\ \text{in})\left(\frac{1\ \text{kip}}{1000\ \text{lbf}}\right) \\ &= 26.29\ \text{kips} \end{aligned}$$

$$\frac{\phi V_c}{2} = \frac{(0.85)(26.29\ \text{kips})}{2} = 11.2\ \text{kips}$$

Since $V_u > \phi V_c/2$, shear reinforcement is required.

The ACI minimum required shear reinforcement for a stirrup spacing, s, of 12 in is

$$A_v = \frac{50 b s}{f_y} = \frac{(50)(12\ \text{in})(12\ \text{in})}{40{,}000\ \frac{\text{lbf}}{\text{in}^2}} = 0.18\ \text{in}^2$$

The nominal shear strength provided by reinforcement is given by $V_s = A_v f_y d/s$, and the amount of shear reinforcement, A_v, required by this equation can be found by using $\phi(V_c + V_s) \geq V_u$ and rearranging to solve for A_v as follows.

$$\begin{aligned} A_v &= \frac{V_u - \phi V_c}{(\phi f_y)\left(\frac{d}{s}\right)} \\ &= \frac{32.5\ \text{kips} - (0.85)(26.29\ \text{kips})}{(0.85)\left(40{,}000\ \frac{\text{lbf}}{\text{in}^2}\right)\left(\frac{1\ \text{kip}}{1000\ \text{lbf}}\right)\left(\frac{20\ \text{in}}{12\ \text{in}}\right)} \\ &= 0.179\ \text{in}^2 \end{aligned}$$

The larger value for A_v controls. Therefore, $A_v = 0.18\ \text{in}^2$.

(Note that although not required for this problem, a check should be made to ensure that V_s does not exceed the ACI-allowed maximum shear reinforcement given by $V_{s(\max)} = 8\sqrt{f'_c}\, bd$ in an actual design and analysis situation.)

Answer is C.

39. A square column is supported by a square reinforced concrete footing with depth to reinforcement d = 33 in as shown. The column supports a dead load of 200 kips and a live load of 100 kips. The American Concrete Institute (ACI) code requires that the loaded area of footing for beam shear starts at a distance of d away from the column face and that the loaded area of footing for punching shear starts at a distance of $d/2$ away from the column face.

b = 8 ft
a = 18 in
18 in
8 ft

column load: D = 200 kips
L = 100 kips
d = 33 in
36 in
reinforcing steel centroid

In accordance with ACI strength design, the controlling (maximum) factored shear stress is most nearly

(A) 25 lbf/in^2
(B) 30 lbf/in^2
(C) 35 lbf/in^2
(D) 48 lbf/in^2

Solution:

The ultimate applied load is

$$\begin{aligned} P_u &= 1.4D + 1.7L \\ &= (1.4)(200\ \text{kips}) + (1.7)(100\ \text{kips}) \\ &= 450\ \text{kips} \end{aligned}$$

The net ultimate soil pressure, q_u, is given by

$$q_u = \frac{P_u}{A} = \frac{450\ \text{kips}}{(8\ \text{ft})(8\ \text{ft})} = 7.0\ \text{kips/ft}^2$$

Check beam shear.

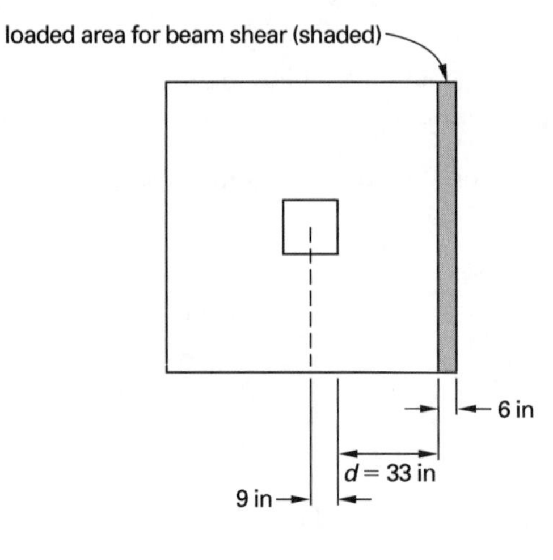

The factored shear stress for beam shear is

$$\begin{aligned} v_u &= \frac{V_u}{bd} \\ &= \frac{\left(7.0\ \frac{\text{kips}}{\text{ft}^2}\right)\left(\frac{6\ \text{in}}{12\ \frac{\text{in}}{\text{ft}}}\right)(8\ \text{ft})\left(1000\ \frac{\text{lbf}}{\text{kips}}\right)}{(8\ \text{ft})\left(12\ \frac{\text{in}}{\text{ft}}\right)(33\ \text{in})} \\ &= 8.8\ \text{lbf/in}^2 \end{aligned}$$

Check punching shear.

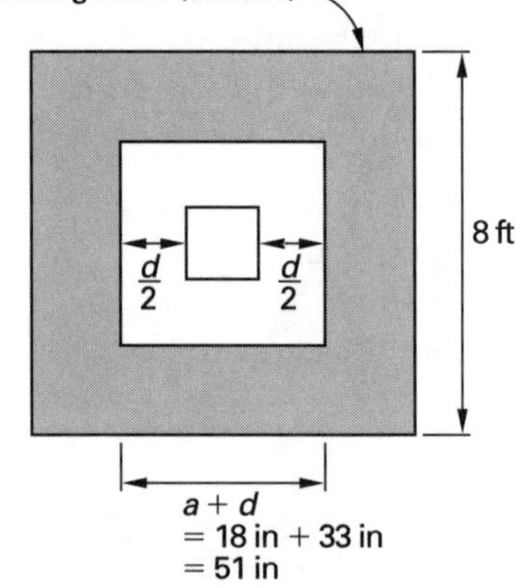

The factored shear stress for punching shear is

$$v_u = \frac{V_u}{(4)(a+d)d}$$

$$= \frac{\left(7.0\ \frac{\text{kips}}{\text{ft}^2}\right)\left[(8\ \text{ft})(8\ \text{ft}) - \frac{(51\ \text{in})(51\ \text{in})}{144\ \frac{\text{in}^2}{\text{ft}^2}}\right]\left(1000\ \frac{\text{lbf}}{\text{kips}}\right)}{(4)(51\ \text{in})(33\ \text{in})}$$

$$= 47.8\ \text{lbf/in}^2 \quad (48\ \text{lbf/in}^2)$$

The punching shear stress is larger than the beam shear stress. Therefore, punching shear controls.

Answer is D.

Problems 40–42 are based on the following information and illustration.

A floor system consists of 20 reinforced concrete beams and a continuous 3 in deck slab. (A typical section is shown for the deck and two of the beams.) Assume the beams are underreinforced.

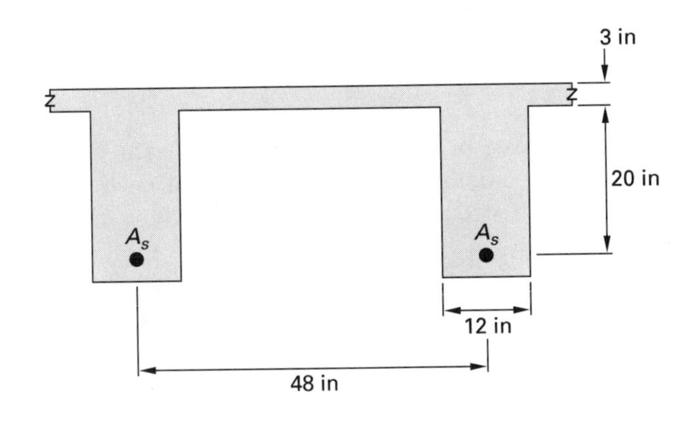

$f'_c = 3000\ \text{lbf/in}^2$
$f_y = 60{,}000\ \text{lbf/in}^2$
$L = 30\ \text{ft}$ [simple span length]

40. For each beam in the floor system, the ACI-specified effective top flange width is most nearly

(A) 15 in
(B) 48 in
(C) 60 in
(D) 90 in

Solution:

The effective flange width, b_e, is given by

$$b_e = \min \begin{cases} \left(\frac{1}{4}\right)(\text{beam span}) = \left(\frac{1}{4}\right)(30\ \text{ft})\left(\frac{12\ \text{in}}{1\ \text{ft}}\right) \\ \quad = 90\ \text{in} \\ b_w + (16)(\text{slab depth}) = 12\ \text{in} + (16)(3\ \text{in}) \\ \quad = 60\ \text{in} \\ b_w + \begin{matrix}\text{clear span}\\ \text{between beams}\end{matrix} = 12\ \text{in} + 36\ \text{in} \\ \quad = 48\ \text{in} \end{cases}$$

Therefore, $b_e = 48$ in.

Answer is B.

41. If the area of reinforcing steel per beam, A_s, is 7.25 in^2, the nominal moment capacity of each beam based on ACI strength design is most nearly

(A) 680 ft-kips
(B) 770 ft-kips
(C) 800 ft-kips
(D) 880 ft-kips

Solution:

This problem asks for the *nominal* moment capacity, M_n, not the *allowable* moment capacity, ϕM_n. Therefore, use of the reduction factor, ϕ, does not apply here.

The depth of the concrete compressive stress block must be checked to see whether or not it exceeds the 3 in deck thickness. If the depth of this compressive stress block exceeds the deck thickness, then it is a T-beam and T-beam formulas apply for determination of the nominal moment capacity. If, however, the depth of the concrete compressive stress block does not exceed the deck thickness, then it is a rectangular beam and rectangular beam formulas apply for determination of the nominal moment capacity.

First, assuming that each beam is rectangular with a width, $b = b_e = 48$ in, the depth of the concrete compressive stress block, a, is given by

$$a = \frac{A_s f_y}{0.85 f'_c b} = \frac{(7.25\ \text{in}^2)\left(60{,}000\ \frac{\text{lbf}}{\text{in}^2}\right)}{(0.85)\left(3000\ \frac{\text{lbf}}{\text{in}^2}\right)(48\ \text{in})} = 3.55\ \text{in}$$

Since $a >$ slab depth of 3 in, the beam is a T-beam.

Since $a = 3.55$ in was found by assuming that the beam was a rectangular beam with width $b = 48$ in, this a only indicates whether or not the beam is a T-beam and is not the correct a for determining the nominal moment capacity. The correct a is now found by applying the concepts of static equilibrium to the beam.

To find the correct a, sum horizontal forces in the T-beam to show that the upper (above the neutral axis) compressive concrete stress block force is equal to the lower (below the neutral axis) maximum tensile force sustained by the reinforcing bars. By dividing the entire concrete compressive stress block section into three parts (a rectangular part and two overhanging "flanges"), a can be found.

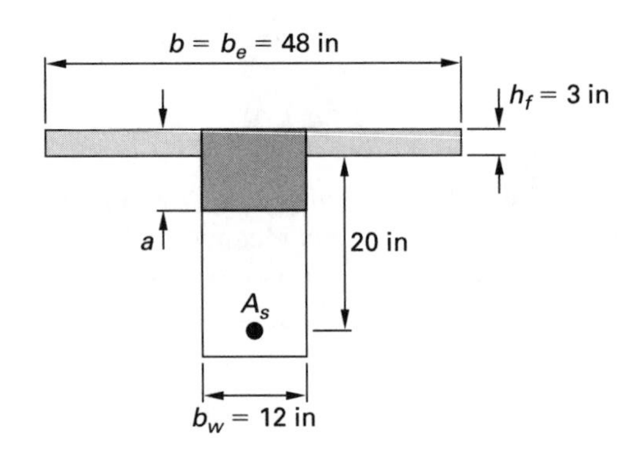

$$\begin{aligned} A_f &= \text{area of overhanging “flanges”} \\ &= (b_e - b_w)h_f \\ &= (48\text{ in} - 12\text{ in})(3\text{ in}) \\ &= 108\text{ in}^2 \end{aligned}$$

A_c = total area of concrete compressive stress block

A_r = area of concrete compressive stress block in the rectangular part of the T-beam between the overhanging "flanges"

From equilibrium of horizontal forces,

$$0.85 f'_c A_c = A_s f_y$$

$$A_c = \frac{A_s f_y}{0.85 f'_c} = \frac{(7.25\text{ in}^2)\left(60{,}000\ \frac{\text{lbf}}{\text{in}^2}\right)}{(0.85)\left(3000\ \frac{\text{lbf}}{\text{in}^2}\right)} = 170.59\text{ in}^2$$

$$A_r = b_w a = A_c - A_f$$

$$a = \frac{A_c - A_f}{b_w} = \frac{170.59\text{ in}^2 - 108\text{ in}^2}{12\text{ in}} = 5.22\text{ in}$$

The nominal moment capacity of the T-beam, M_n, is found by

$$\begin{aligned} M_n &= 0.85 f'_c h_f (b_e - b_w)\left(d - \frac{h_f}{2}\right) + 0.85 f'_c a b_w \left(d - \frac{a}{2}\right) \\ &= \left[\begin{array}{l} (0.85)\left(3000\ \frac{\text{lbf}}{\text{in}^2}\right)(3\text{ in}) \\ \quad \times (48\text{ in} - 12\text{ in})\left(23\text{ in} - \frac{3\text{ in}}{2}\right) \\ + (0.85)\left(3000\ \frac{\text{lbf}}{\text{in}^2}\right)(5.22\text{ in}) \\ \quad \times (12\text{ in})\left(23\text{ in} - \frac{5.22\text{ in}}{2}\right) \end{array}\right] \\ &\quad \times \left(\frac{1\text{ ft}}{12\text{ in}}\right)\left(\frac{1\text{ kip}}{1000\text{ lbf}}\right) \\ &= 765\text{ ft-kips} \quad (770\text{ ft-kips}) \end{aligned}$$

(Note that even though the problem statement allows the assumption that the beam is underreinforced, the actual reinforcing steel ratio and its limits should always be checked in real design/analysis problems.)

Answer is B.

42. If the area of reinforcing steel per beam, A_s, is 6.00 in^2, the nominal moment capacity of each beam based on ACI strength design is most nearly

(A) 150 ft-kips
(B) 160 ft-kips
(C) 590 ft-kips
(D) 650 ft-kips

Solution:

This problem asks for the *nominal* moment capacity, M_n, not the *allowable* moment capacity, ϕM_n. Therefore, use of the reduction factor, ϕ, does not apply here.

First, assuming that each beam is a rectangular beam with a width, $b = b_e = 48$ in, the depth of the concrete compressive stress block, a, is given by

$$a = \frac{A_s f_y}{0.85 f'_c b} = \frac{(6.00\text{ in}^2)\left(60{,}000\ \frac{\text{lbf}}{\text{in}^2}\right)}{(0.85)\left(3000\ \frac{\text{lbf}}{\text{in}^2}\right)(48\text{ in})} = 2.94\text{ in}$$

Since $a <$ slab depth of 3 in, the nominal moment capacity of the beam, M_n, is the same as for a rectangular singly reinforced concrete beam, using $b = b_e =$ 48 in, and is given by

$$M_n = 0.85 f_c' a b \left(d - \frac{a}{2}\right)$$

$$= \left[\begin{array}{l}(0.85)\left(3000\ \frac{\text{lbf}}{\text{in}^2}\right)(2.94\ \text{in}) \\ \times (48\ \text{in})\left(23\ \text{in} - \frac{2.94\ \text{in}}{2}\right)\end{array}\right]$$

$$\times \left(\frac{1\ \text{ft}}{12\ \text{in}}\right)\left(\frac{1\ \text{kip}}{1000\ \text{lbf}}\right)$$

$$= 646\ \text{ft-kips}\quad (650\ \text{ft-kips})$$

This can also be calculated by using $M_n = A_s f_y (d - a/2)$ in accordance with rectangular singly reinforced concrete beam theory.

(Note that even though the problem statement allows the assumption that the beam is underreinforced, the actual reinforcing steel ratio and its limits should always be checked in real design/analysis problems.)

Answer is D.

SURVEYING

Problems 43 and 44 are based on the following information.

Earthwork quantities for a section of roadway indicate a transition from fill to cut. The following areas are scaled from the print cross sections.

station (m)	cut area (m^2)	fill area (m^2)
20+00		173.21
20+10.50		43.56
20+21.50	14.32	9.63
20+28.45	64.73	
20+40	187.42	

In the region where there is a transition from fill to cut, the fill area and cut area are both triangular in shape on the road cross section.

43. The total volume of fill required for this section of road is most nearly

(A) 1430 m^3
(B) 1450 m^3
(C) 1730 m^3
(D) 1780 m^3

Solution:

Earthwork volumes for fill areas and cut areas can be calculated using the average end area formula. Since the cut and fill areas are triangular in shape as given in the problem statement, earthwork volumes in the transition region from fill to cut can be calculated from the formula that gives the volume of a pyramid.

sta 20+00 to 20+10.50:

$$L = 10.50\ \text{m} - 0\ \text{m} = 10.50\ \text{m}$$

$$\text{fill volume} = L\left(\frac{A_1 + A_2}{2}\right)$$

$$= (10.50\ \text{m})\left(\frac{173.21\ \text{m}^2 + 43.56\ \text{m}^2}{2}\right)$$

$$= 1138.0\ \text{m}^3$$

sta 20+10.50 to 20+21.50:

$$L = 21.50\ \text{m} - 10.50\ \text{m} = 11.00\ \text{m}$$

(This is the transition from fill to cut, so use the formula for pyramid volume to calculate cut area.)

$$\text{fill volume} = L\left(\frac{A_1 + A_2}{2}\right)$$

$$= (11.00\ \text{m})\left(\frac{43.56\ \text{m}^2 + 9.63\ \text{m}^2}{2}\right)$$

$$= 292.5\ \text{m}^3$$

$$\text{cut volume} = h\left(\frac{\text{area of base}}{3}\right)$$

$$= (11.00\ \text{m})\left(\frac{14.32\ \text{m}^2}{3}\right)$$

$$= 52.5\ \text{m}^3$$

sta 20+21.50 to 20+28.45:

$$L = 28.45\ \text{m} - 21.50\ \text{m} = 6.95\ \text{m}$$

(This is the transition from fill to cut, so use the formula for pyramid volume to calculate fill area.)

$$\text{fill volume} = h\left(\frac{\text{area of base}}{3}\right)$$

$$= (6.95\ \text{m})\left(\frac{9.63\ \text{m}^2}{3}\right)$$

$$= 22.3\ \text{m}^3$$

$$\text{cut volume} = L\left(\frac{A_1 + A_2}{2}\right)$$

$$= (6.95\ \text{m})\left(\frac{14.32\ \text{m}^2 + 64.73\ \text{m}^2}{2}\right)$$

$$= 274.7\ \text{m}^3$$

sta 20+28.45 to 20+40:

$$L = 40\text{ m} - 28.45\text{ m} = 11.55\text{ m}$$

$$\begin{aligned}\text{cut volume} &= L\left(\frac{A_1 + A_2}{2}\right)\\ &= (11.55\text{ m})\left(\frac{64.73\text{ m}^2 + 187.42\text{ m}^2}{2}\right)\\ &= 1456.2\text{ m}^3\end{aligned}$$

A table that summarizes earthwork volumes is now made.

station (m)	cut area (m²)	fill area (m²)	cut volume (m³)	fill volume (m³)
20+00		173.21		
				1138.0
20+10.50		43.56		
			52.5	292.5
20+21.50	14.32	9.63		
			274.7	22.3
20+28.45	64.73			
			1456.2	
20+40	187.42			
		total	1783.4	1452.8

Therefore, the total volume of fill required for this section of road is 1452.8 m^3 (1450 m^3).

Answer is B.

44. The total volume of cut required for this section of road is most nearly

(A) 1430 m^3
(B) 1453 m^3
(C) 1731 m^3
(D) 1783 m^3

Solution:

From Prob. 43, the total volume of cut required for this section of road is 1783.4 m^3 (1780 m^3).

Answer is D.

Problems 45–47 are based on the following information and illustration.

The proposed vertical profile for a transport airport runway is shown.

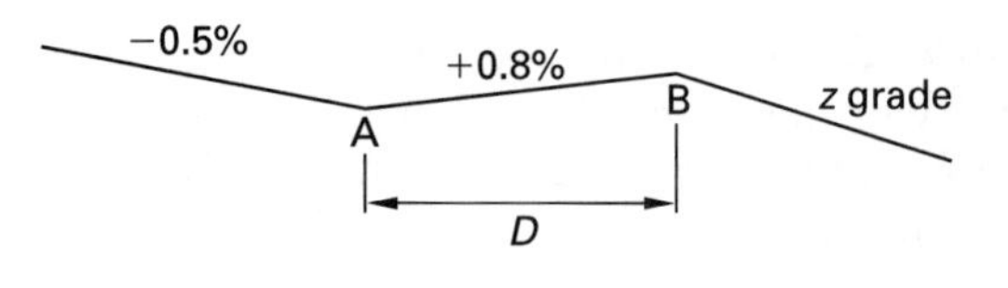

45. The minimum required length of vertical curve at the point of vertical intersection A is most nearly

(A) 300 ft
(B) 390 ft
(C) 1000 ft
(D) 1300 ft

Solution:

At this point, the grade change is from −0.5% to +0.8%. Therefore, the absolute value of the total percent grade change at the point of vertical intersection A is

$$\Delta_A = |-0.5\% - (+0.8\%)| = 1.3\%$$

For transport airports, the minimum required length of vertical curve at the point of vertical intersection A is

$$\begin{aligned}L_A &= \left(\frac{1000\text{ ft}}{1\%\text{ change}}\right)(\Delta_A)\\ &= \left(\frac{1000\text{ ft}}{1\%\text{ change}}\right)(1.3\%\text{ change})\\ &= 1300\text{ ft}\end{aligned}$$

Answer is D.

46. The maximum allowed longitudinal grade (maximum allowed downward slope), z, to the right of the point of vertical intersection B is most nearly

(A) −2.0%
(B) −1.5%
(C) −1.2%
(D) −0.7%

Solution:

Two criteria must be checked for this problem. First, the maximum allowed longitudinal grade change is 1.5% for transport airports. Second, the maximum longitudinal grade is 1.5% for transport airports.

Checking the first criteria, the absolute value of the total grade change at the point of vertical intersection B can be rearranged to give the maximum allowed longitudinal grade (maximum allowed downward slope), z, as being

$$\begin{aligned}\Delta_{\text{maxB}} &= 1.5\% = +0.8\% - z\\ z &= 0.8\% - 1.5\% = -0.7\%\end{aligned}$$

The second criteria is that $|z|$ not exceed 1.5%, which it does not. Therefore, the maximum allowed longitudinal grade (maximum allowed downward slope), z, is −0.7%.

Answer is D.

47. If the grade to the right of the point of vertical intersection B is −0.4% (i.e., $z = -0.4\%$), the minimum required distance, D, between points of vertical intersection A and B is most nearly

(A) 250 ft
(B) 630 ft
(C) 1000 ft
(D) 2500 ft

Solution:

From Prob. 45, the grade change at the point of vertical intersection A is $\Delta_A = 1.3\%$. For a −0.4% grade at the point of vertical intersection B,

$$\Delta_B = |+0.8\% - (-0.4\%)| = 1.2\%$$

Therefore, for transport airports, the minimum required distance between the points of vertical intersection A and B is

$$\begin{aligned} D &= \left(1000\ \frac{\text{ft}}{\%}\right)(\Delta_A + \Delta_B) \\ &= \left(1000\ \frac{\text{ft}}{\%}\right)(1.3\% + 1.2\%) \\ &= 2500\ \text{ft} \end{aligned}$$

Answer is D.

48. A boundary and traverse line are shown.

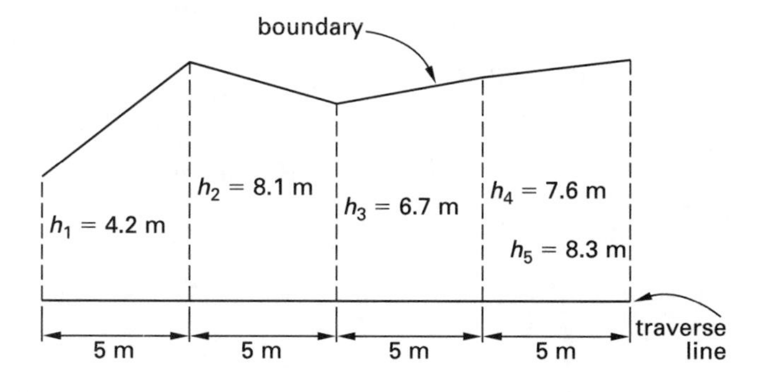

Using the trapezoidal rule, the total area between the curved boundary and traverse line is most nearly

(A) 141 m^2
(B) 143 m^2
(C) 148 m^2
(D) 151 m^2

Solution:

By the trapezoidal rule, the area is

$$\begin{aligned} & w\left(\frac{h_1 + h_5}{2} + h_2 + h_3 + h_4\right) \\ &= (5\ \text{m})\left(\frac{4.2\ \text{m} + 8.3\ \text{m}}{2} + 8.1\ \text{m} + 6.7\ \text{m} + 7.6\ \text{m}\right) \\ &= 143.2\ \text{m}^2 \quad (143\ \text{m}^2) \end{aligned}$$

Answer is B.

TRANSPORTATION

Problems 49 and 50 are based on the following information.

The connection matrix shown represents a road transportation network between 6 locations.

node	1	2	3	4	5	6
1	0	1	0	0	1	1
2	1	0	1	0	0	1
3	0	1	0	1	1	0
4	0	0	1	0	1	0
5	1	0	−1	1	0	1
6	1	1	0	0	1	0

49. The total number of links in the network is

(A) 7
(B) 8
(C) 9
(D) 10

Solution:

By connecting the nodes in accordance with the connection matrix, a graphical representation of the road transportation network can be seen.

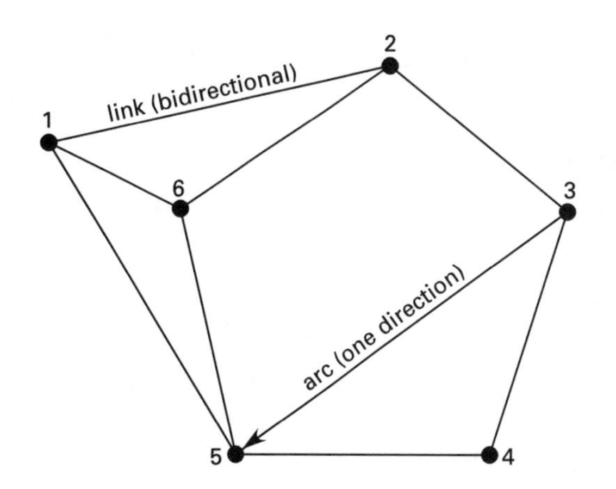

A link is defined by the nodes that exist at both its ends and does not specify direction. Therefore, there are eight links in this network.

Answer is B.

50. The total number of arcs in the network is

(A) 0
(B) 1
(C) 2
(D) 3

Solution:

An arc is a link with a specific direction assigned to it. From the graphical representation of the network shown in Prob. 49, there is one arc. In the connection matrix, a negative number indicates that the direction of travel from one node to another goes against the direction assigned to that arc.

Answer is B.

Problems 51 and 52 are based on the following information.

A traffic flow relationship is given by $q = k\mathrm{v}$, where $q =$ traffic volume in veh/hr, $k =$ traffic density in veh/mi, and $\mathrm{v} =$ mean speed in mi/hr. The mean speed on a road in mi/hr is given by the relationship $\mathrm{v} = 60 - 0.2k$.

51. If the mean speed on a road during the rush hour is 45 mi/hr, the maximum capacity of traffic density for this road during rush hour is most nearly

(A) 15 veh/mi
(B) 45 veh/mi
(C) 75 veh/mi
(D) 230 veh/mi

Solution:

The mean velocity relationship is rearranged to give

$$k = \frac{60 - \mathrm{v}}{0.2} = \frac{60\ \frac{\text{mi}}{\text{hr}} - 45\ \frac{\text{mi}}{\text{hr}}}{0.2\ \frac{\text{mi}^2}{\text{veh}\cdot\text{hr}}} = 75\ \text{veh/mi}$$

Answer is C.

52. The maximum capacity of overall traffic volume for this road is most nearly

(A) 3400 veh/hr
(B) 4300 veh/hr
(C) 4500 veh/hr
(D) 5000 veh/hr

Solution:

The mean speed relationship can be substituted into the traffic flow relationship resulting in a quadratic relationship (i.e., a parabolic curve).

$$q = k\mathrm{v} = k(60 - 0.2k) = 60k - 0.2k^2$$

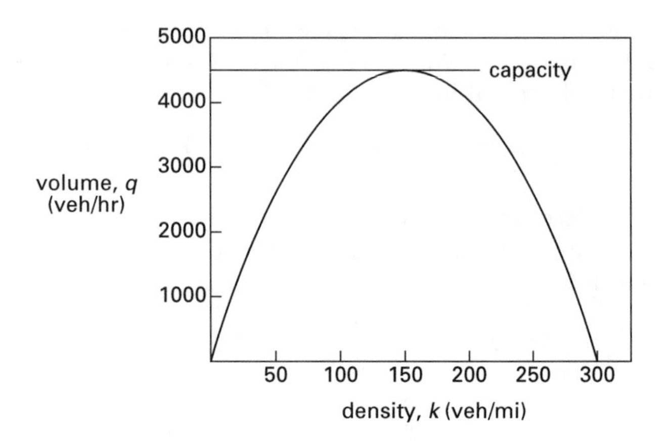

To determine the traffic volume capacity, it is necessary to find the maximum point on the parabolic curve (i.e., the location where the slope of the curve equals zero).

$$\frac{dq}{dk} = 0$$

$$\frac{d(60k - 0.2k^2)}{dk} = 60 - 0.4k = 0$$

$$k = \frac{60}{0.4} = 150\ \text{veh/mi}$$

Substituting $k = 150$ veh/mi into the traffic flow relationship gives

$$\begin{aligned} q &= 60k - 0.2k^2 \\ &= (60)\left(150\ \frac{\text{veh}}{\text{hr}}\right) - (0.2)\left(150\ \frac{\text{veh}}{\text{hr}}\right)^2 \\ &= 4500\ \text{veh/hr} \end{aligned}$$

Answer is C.

53. A train uses diesel locomotives. Each locomotive is rated at 3500 hp with a diesel-electric efficiency of 82%, has four axles, and weighs 150 tons. Each of the 25 four-axle passenger cars weighs 60 tons and must be pulled up a 2% grade with a long curve that has a 2° curvature. The resistance caused by the 2% gradient can be approximated by $F_g = T(p/100)$, where F_g = gradient force, T = vehicle weight in tons, and p = percent grade. Assume that the wind speed is 0 mi/hr.

The minimum number of locomotives that are required to maintain a speed of 50 mi/hr up the incline is

(A) 2
(B) 3
(C) 4
(D) 5

Solution:

The level tangent resistance (lbf per ton of car weight) of one passenger car can be found by the modified Davis equation (using $K = 0.07$ for standard rail units) to be

$$\begin{aligned} R &= 0.6 + \frac{20}{W} + 0.01V + \frac{KV^2}{WN} \\ &= 0.6 + \frac{20}{\frac{60 \text{ tons}}{4 \text{ axles}}} + (0.01)\left(50 \ \frac{\text{mi}}{\text{hr}}\right) \\ &\quad + \frac{(0.07)\left(50 \ \frac{\text{mi}}{\text{hr}}\right)^2}{\left(\frac{60 \text{ tons}}{4 \text{ axles}}\right)(4 \text{ axles})} \\ &= 5.35 \text{ lbf/ton} \end{aligned}$$

Therefore, the total level tangent resistance, F_R, for one passenger car is

$$F_R = RT = \left(5.35 \ \frac{\text{lbf}}{\text{ton}}\right)\left(60 \ \frac{\text{tons}}{\text{car}}\right) = 321 \text{ lbf/car}$$

The resistance per passenger car from the 2° track curve is

$$R_c = \left(0.8 \ \frac{\text{lbf}}{\text{ton}}\right)(60 \text{ tons})(2°) = 96 \text{ lbf/car}$$

The resistance per passenger car from the 2% gradient is

$$\begin{aligned} F_g &= T\left(\frac{p}{100}\right) = (60 \text{ tons})\left(\frac{2000 \text{ lbf}}{1 \text{ ton}}\right)\left(\frac{2\%}{100}\right) \\ &= 2400 \text{ lbf/car} \end{aligned}$$

The level tangent resistance (lbf per ton of locomotive weight) of one locomotive can be found by the modified Davis equation to be

$$\begin{aligned} R &= 0.6 + \frac{20}{W} + 0.01V + \frac{KV^2}{WN} \\ &= 0.6 + \frac{20}{\frac{150 \text{ tons}}{4 \text{ axles}}} + (0.01)\left(50 \ \frac{\text{mi}}{\text{hr}}\right) \\ &\quad + \frac{(0.07)\left(50 \ \frac{\text{mi}}{\text{hr}}\right)^2}{\left(\frac{150 \text{ tons}}{4 \text{ axles}}\right)(4 \text{ axles})} \\ &= 2.8 \text{ lbf/ton} \end{aligned}$$

Therefore, the total level tangent resistance, F_R, for one locomotive is

$$\begin{aligned} F_R = RT &= \left(2.8 \ \frac{\text{lbf}}{\text{ton}}\right)\left(150 \ \frac{\text{tons}}{\text{car}}\right) \\ &= 420 \text{ lbf/locomotive} \end{aligned}$$

The resistance per locomotive from the 2° track curve is

$$R_c = \left(0.8 \ \frac{\text{lbf}}{\text{ton}}\right)(150 \text{ tons})(2°) = 240 \text{ lbf/locomotive}$$

The resistance per locomotive from the 2% gradient is

$$\begin{aligned} F_g &= T\left(\frac{p}{100}\right) = (150 \text{ tons})\left(\frac{2000 \text{ lbf}}{1 \text{ ton}}\right)\left(\frac{2\%}{100}\right) \\ &= 6000 \text{ lbf/locomotive} \end{aligned}$$

For a train consisting of n locomotives and 25 passenger cars, the total resistance, R_{total}, is found by summing up the individual resistances, which gives

$$\begin{aligned} R_{\text{total}} &= (25 \text{ cars})\left(321 \ \frac{\text{lbf}}{\text{car}} + 96 \ \frac{\text{lbf}}{\text{car}} + 2400 \ \frac{\text{lbf}}{\text{car}}\right) \\ &\quad + n\left(\begin{array}{c} 420 \ \frac{\text{lbf}}{\text{locomotive}} + 240 \ \frac{\text{lbf}}{\text{locomotive}} \\ + 6000 \ \frac{\text{lbf}}{\text{locomotive}} \end{array}\right) \\ &= 70{,}425 \text{ lbf} + n\left(6660 \ \frac{\text{lbf}}{\text{locomotive}}\right) \end{aligned}$$

Now calculate the tractive effort of each locomotive. One locomotive produces a tractive effort of

$$\begin{aligned} \text{TE} &= \frac{375(\text{HP})e}{V} = \frac{(375)(3500 \text{ hp})(0.82)}{50 \ \frac{\text{mi}}{\text{hr}}} \\ &= 21{,}525 \text{ lbf/locomotive} \end{aligned}$$

The total tractive effort for n locomotives is

$$\text{TE}_{\text{total}} = n\left(21{,}525\ \frac{\text{lbf}}{\text{locomotive}}\right)$$

The total minimum number of locomotives required can be found by setting TE_{total} equal to R_{total}.

$$\text{TE}_{\text{total}} = R_{\text{total}}$$

$$n\left(21{,}525\ \frac{\text{lbf}}{\text{locomotive}}\right) = 70{,}425\ \text{lbf} + n\left(6660\ \frac{\text{lbf}}{\text{locomotive}}\right)$$

Solving for n,

$$n = \frac{70{,}425\ \text{lbf}}{21{,}525\ \frac{\text{lbf}}{\text{locomotive}} - 6660\ \frac{\text{lbf}}{\text{locomotive}}} = 4.7\ \text{locomotives}\quad (5\ \text{locomotives})$$

Answer is D.

54. A one-lane rural road has a 10° curve extending for 230 m along its centerline. The road is 5 m wide with 3 m wide shoulders. The design speed for this road is 75 km/h.

The superelevation needed on the curve to give a side-friction factor of zero at the design speed is most nearly

(A) 0.0005
(B) 0.034
(C) 1.1
(D) 1.9

Solution:

The radius of curvature is

$$R = \frac{s}{\phi} = \frac{230\ \text{m}}{(10^\circ)\left(\frac{\pi\ \text{rad}}{180^\circ}\right)} = 1318\ \text{m}$$

For a side-friction factor of $f = 0$, the superelevation is given by

$$e + f = e = \frac{\text{v}^2}{gR} = \frac{\left[\left(75\ \frac{\text{km}}{\text{h}}\right)\left(\frac{1000\ \text{m}}{1\ \text{km}}\right)\left(\frac{1\ \text{h}}{3600\ \text{s}}\right)\right]^2}{\left(9.81\ \frac{\text{m}}{\text{s}^2}\right)(1318\ \text{m})} = 0.0336\quad (0.034)$$

Answer is B.

WATER PURIFICATION AND TREATMENT

55. A water treatment plant adds chlorine gas at a daily concentration of 1.5 mg/L. The volume of water treated each day is 25,000,000 gal.

The daily amount of chlorine required is most nearly

(A) 38 kg
(B) 64 kg
(C) 95 kg
(D) 142 kg

Solution:

The total daily amount of chlorine required is

$$\left(1.5\ \frac{\text{mg}}{\text{L}}\right)\left(\frac{1\ \text{g}}{1000\ \text{mg}}\right)\left(\frac{1\ \text{kg}}{1000\ \text{g}}\right) \times \left(\frac{3.785\ \text{L}}{1\ \text{gal}}\right)(25{,}000{,}000\ \text{gal}) = 142\ \text{kg}$$

Answer is D.

56. A gas chlorination system at a water treatment plant feeds 23 lbm of chlorine per day into a daily water flow of 5,000,000 gal.

If the daily chlorine demand is 0.5 mg/L, the daily chlorine residual is most nearly

(A) 0.02 mg/L
(B) 0.05 mg/L
(C) 0.08 mg/L
(D) 0.6 mg/L

Solution:

The total amount of chlorine per unit volume of water is

$$\left(\frac{23\ \frac{\text{lbm}}{\text{day}}}{5{,}000{,}000\ \frac{\text{gal}}{\text{day}}}\right)\left(\frac{0.454\ \text{kg}}{1\ \text{lbm}}\right) \times \left(\frac{1000\ \text{g}}{1\ \text{kg}}\right)\left(\frac{1000\ \text{mg}}{1\ \text{g}}\right)\left(\frac{1\ \text{gal}}{3.785\ \text{L}}\right) = 0.552\ \text{mg/L}$$

The daily chlorine demand is 0.5 mg/L.

The daily chlorine residual is

$$0.552\ \frac{\text{mg}}{\text{L}} - 0.5\ \frac{\text{mg}}{\text{L}} = 0.052\ \text{mg/L}\quad (0.05\ \text{mg/L})$$

Answer is B.

57. The five-day BOD of a wastewater sample is 226 mg/L. The ultimate BOD is 319 mg/L.

The BOD reaction rate constant (base e) is most nearly

(A) 0.030 days^{-1}
(B) 0.069 days^{-1}
(C) 0.11 days^{-1}
(D) 0.25 days^{-1}

Solution:

Rearranging the equation for BOD exertion gives

$$y_t = L(1 - e^{-kt})$$

$$k = \frac{\ln\left(1 - \frac{y_t}{L}\right)}{-t} = \frac{\ln\left(1 - \frac{226\ \frac{\text{mg}}{\text{L}}}{319\ \frac{\text{mg}}{\text{L}}}\right)}{-(5\ \text{days})}$$

$$= 0.247\ \text{days}^{-1} \quad (0.25\ \text{days}^{-1})$$

Answer is D.

Problems 58 and 59 are based on the following information.

A wastewater sample has a five-day BOD of 250 mg/L at 20°C and a BOD reaction rate constant of 0.15 days^{-1} (base e) at 20°C.

The BOD reaction rate constant, k, varies with temperature in accordance with $k_T = k_{20}1.047^{T-20}$, where k_{20} is the BOD reaction rate constant at a temperature of 20°C.

58. The ultimate BOD demand is most nearly

(A) 250 mg/L
(B) 300 mg/L
(C) 470 mg/L
(D) 530 mg/L

Solution:

The ultimate BOD can be calculated from the data at 20°C.

$$y_t = L(1 - e^{-kt})$$

$$L = \frac{y_t}{1 - e^{-kt}} = \frac{250\ \frac{\text{mg}}{\text{L}}}{1 - e^{-(0.15\ \text{days}^{-1})(5\ \text{days})}}$$

$$= 473.8\ \text{mg/L} \quad (470\ \text{mg/L})$$

Answer is C.

59. The 10-day BOD at a temperature of 18°C is most nearly

(A) 190 mg/L
(B) 230 mg/L
(C) 350 mg/L
(D) 390 mg/L

Solution:

From Prob. 38, the ultimate BOD is $L = 473.8$ mg/L. The BOD reaction rate constant at 18°C is

$$k_{18} = k_{20}1.047^{T-20} = (0.15\ \text{days}^{-1})(1.047)^{18-20}$$
$$= 0.137\ \text{days}^{-1}$$

The 10-day BOD at 18°C is

$$y_t = L(1 - e^{-kt})$$
$$= \left(473.8\ \frac{\text{mg}}{\text{L}}\right)\left(1 - e^{-(0.137\ \text{days}^{-1})(10\ \text{days})}\right)$$
$$= 353.4\ \text{mg/L} \quad (350\ \text{mg/L})$$

Answer is C.

60. A trickling filter plant treats 1000 m^3/day of sewage sludge with a BOD_5 of 200 mg/L and suspended solids of 230 mg/L. Primary clarification removes 30% of the BOD and 60% of the influent solids. The solids generated in the secondary processes average 0.4 kg/kg BOD_5.

The total solids production is most nearly

(A) 56 kg/day
(B) 140 kg/day
(C) 190 kg/day
(D) 230 kg/day

Solution:

Removal in the primary processes is

$$(0.6)\left(230\ \frac{\text{mg}}{\text{L}}\right) = 138\ \text{mg/L}$$

Production in the secondary processes is

$$(1 - 0.3)\left(200\ \frac{\text{mg}}{\text{L}}\right)\left(0.4\ \frac{\text{kg}}{\text{kg}}\right) = 56\ \text{mg/L}$$

The total solids production is

$$\left(138\ \frac{\text{mg}}{\text{L}} + 56\ \frac{\text{mg}}{\text{L}}\right)\left(\frac{1\ \text{g}}{1000\ \text{mg}}\right) \times \left(\frac{1\ \text{kg}}{1000\ \text{g}}\right)\left(\frac{1000\ \text{L}}{1\ \text{m}^3}\right)\left(1000\ \frac{\text{m}^3}{\text{day}}\right) = 194\ \text{kg/day}\quad (190\ \text{kg/day})$$

Answer is C.

Practice Exam

PROBLEMS FOR THE PRACTICE EXAM

1. The function $y = f(x) = e^x/(e+x)^x$ describes the placement of a line of trees with respect to three straight-line property boundaries as shown. There is a clearing between the line of trees and the property boundaries.

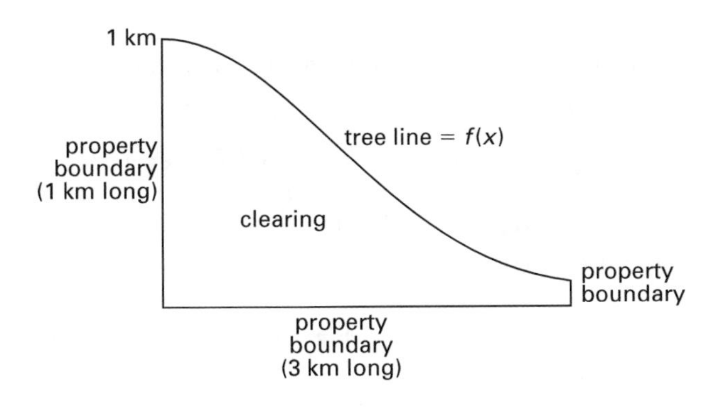

Using Simpson's 1/3 rule and a horizontal interval of $\Delta x = 0.5$ km, the area of the clearing is most nearly

(A) 1.5 km^2
(B) 1.6 km^2
(C) 3.3 km^2
(D) 4.9 km^2

Problems 2 and 3 are based on the following information and illustration.

A structural system consists of two beam elements AB and BC as shown. Assume that translational displacements are limited to the x-y plane; that is, there is no translation in the z-direction.

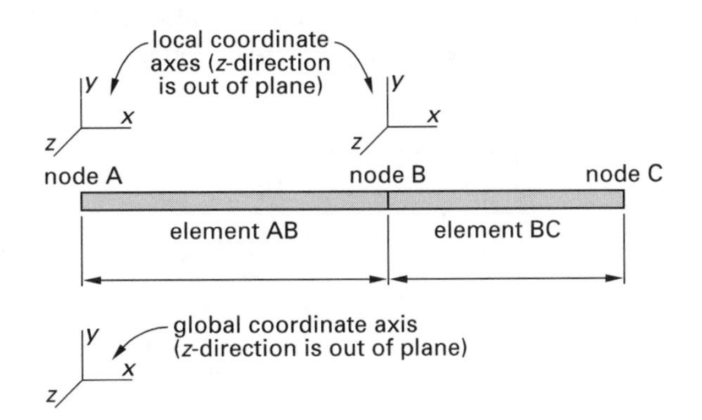

2. The number of degrees of freedom for beam element AB is

(A) 5
(B) 6
(C) 10
(D) 15

3. The number of degrees of freedom for the whole structural system is

(A) 5
(B) 6
(C) 10
(D) 15

Problems 4–6 are based on the following information.

A computer program was written to find the numerical solution to a problem that was also solved analytically.

Assume that methods for determining both the numerical and the analytical solutions were done correctly. The computer program, when executed, went through two iterations to converge on a solution. The first iteration produced a result of 3.110, and the second iteration produced the final answer of 3.132.

The analytical solution produced an answer of 3.1427.

4. The true error with respect to the final solution is most nearly

(A) 0.01
(B) 0.02
(C) 0.3
(D) 0.7

5. The true percent relative error with respect to the final solution is most nearly

(A) 0.01%
(B) 0.02%
(C) 0.3%
(D) 0.7%

6. The approximate percent relative error for the two iterations is most nearly

(A) 0.01%
(B) 0.02%
(C) 0.3%
(D) 0.7%

Problems 7–9 are based on the following information and illustration.

A construction project is composed of activities A through H with the durations, in days, given for each activity in the diagram shown. This project is on a strict schedule that must be maintained and is scheduled to start at the beginning of January 1. Work can only be performed during the day and must be done on every day of the week (Sunday through Saturday).

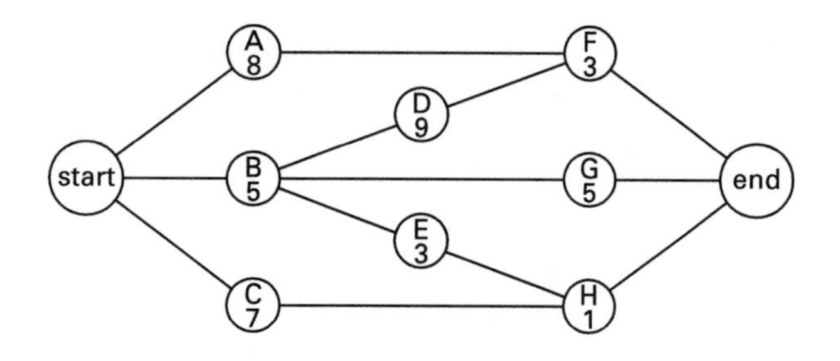

7. The earliest date this project can be completed is

(A) January 8
(B) January 9
(C) January 11
(D) January 17

8. The total float for activity A is

(A) 0 days
(B) 3 days
(C) 6 days
(D) 9 days

9. The latest day that activity E can start is

(A) January 8
(B) January 13
(C) January 14
(D) January 16

10. A 0.2 m layer of soil-bentonite is to be placed beneath the secondary geomembrane liner of a proposed landfill. The layer is to be constructed in two 0.1 m lifts. The bentonite content is to be 5% (dry weight basis). The compacted dry density of the soil/bentonite mixture is 1630 kg/m^3.

The amount of dry bentonite that has to be mixed into each lift is most nearly

(A) 8.2 kg/m^2
(B) 8.6 kg/m^2
(C) 150 kg/m^2
(D) 160 kg/m^2

11. The number of covalent bonds formed by the carbon atom under normal conditions is

(A) 3
(B) 4
(C) 5
(D) 6

12. The number of covalent bonds formed by the nitrogen atom under normal conditions is

(A) 3
(B) 4
(C) 5
(D) 6

Problems 13 and 14 are based on the following information.

A sample of wastewater is incubated for seven days at a temperature of 20°C. After this incubation period, the BOD is found to be 211 mg/L. Assume that the reaction rate constant is 0.14 days^{-1} (base e).

13. The ultimate BOD of this sample is most nearly

(A) 130 mg/L
(B) 210 mg/L
(C) 340 mg/L
(D) 560 mg/L

14. The five-day BOD of this sample is most nearly

(A) 110 mg/L
(B) 170 mg/L
(C) 210 mg/L
(D) 280 mg/L

15. A fresh water sample has a dissolved oxygen concentration of 5.7 mg/L when the temperature is 23.3°C and the atmospheric pressure is 730 mm Hg. A partial listing of the solubility of dissolved oxygen in fresh water at equilibrium with dry air containing 20.9% oxygen and at an atmospheric pressure of 760 mm Hg is as follows.

temperature (°C)	oxygen solubility (mg/L)
21	9.0
22	8.8
23	8.7
24	8.5
25	8.4

The percent saturation of dissolved oxygen in the water sample is most nearly

(A) 63%
(B) 66%
(C) 94%
(D) 96%

16. A reservoir with a water surface at an elevation of 200 m drains through a 1 m diameter pipe with the outlet at an elevation of 180 m. The pipe outlet empties to atmospheric pressure. The total head losses in the pipe and fittings are 18 m. Assume a steady, incompressible flow of 4.92 m^3/s.

A turbine is installed at the pipe outlet. The chosen turbine has an efficiency of 85% and does not add any head loss to the system. The expected power output of the turbine is most nearly

(A) 82 kW
(B) 96 kW
(C) 100 kW
(D) 120 kW

Problems 17 and 18 are based on the following information.

A circular sewer with a 1.5 m inside diameter is designed for a flow rate of 15 m^3/s when flowing full. Assume that the Manning roughness coefficient and Darcy friction factor are variable with depth of flow.

17. The flow rate when the depth of flow is 0.50 m is most nearly

(A) 1.7 m^3/s
(B) 3.4 m^3/s
(C) 5.0 m^3/s
(D) 7.5 m^3/s

18. The velocity of flow when the depth of flow is 0.50 m is most nearly

(A) 2.8 m/s
(B) 4.2 m/s
(C) 4.8 m/s
(D) 6.6 m/s

19. A property consists of 7500 m^2 of lawn area with a runoff coefficient of 0.20, 2000 m^2 of gravel roadway with a runoff coefficient of 0.15, and 500 m^2 of roof surfaces with a runoff coefficient of 0.80.

The overall runoff coefficient for the entire property is most nearly

(A) 0.15
(B) 0.22
(C) 0.38
(D) 0.80

Problems 20 and 21 are based on the following information.

A housing development has a population of 20,000 people. The average sewage flow is 8000 m^3/day.

20. The estimated minimum sewage flow for this housing development is most nearly

(A) 2000 m^3/day
(B) 2600 m^3/day
(C) 4000 m^3/day
(D) 6000 m^3/day

21. The estimated peak sewage flow for this housing development is most nearly

(A) 8000 m^3/day
(B) 20 000 m^3/day
(C) 27 000 m^3/day
(D) 32 000 m^3/day

22. An engineering company strictly adheres to the NCEES "Model Rules of Professional Conduct" for its code of ethics.

According to this code, registered professional engineers have obligations primarily to

(A) society
(B) employers and clients
(C) other registered professional engineers
(D) A, B, and C

Problems 23 and 24 are based on the following information.

(Note that the following description is purely hypothetical and is specific to this problem only.)

The Acme Building Company manufactured and sold a prefabricated steel frame building, designed by its own engineers and architects, to one of its local dealers, East Side Construction Company, which then erected the building as a plant facility in 1980 for a customer.

In 1986, the building collapsed from ponding of water on the roof in a light rainfall. Both Acme and the building owner retained forensic engineers for the subsequent investigation, who concluded that the building roof was adequately designed to support uniform live loads but not unbalanced live loads.

Prevailing design codes stipulate that all loading conditions producing maximum stresses and maximum deflections must be accounted for in structural designs. However, several other prefabricated steel building manufacturers in the region were contacted who testified that common industry practice is to design the roofs on these types of buildings to sustain only uniform live loads since designing for unbalanced live loads would result in the need for more structural material and make their prefabricated buildings too costly to be competitive with the buildings of other manufacturers. All of these building manufacturers stated that the collapsed Acme building structural design fully satisfied common industry practices in their region.

All engineers involved were registered professional engineers.

23. The engineer of record is most likely to be

(A) Acme Building Company
(B) East Side Construction Company
(C) the building owner
(D) the forensic engineers retained by Acme and the owner

24. Liability for the building collapse lies with

(A) Acme Building Company
(B) East Side Construction Company
(C) the building owner for not regularly inspecting gutters and maintaining proper drainage of water from the roof
(D) Nobody is liable for the collapse since the structural design was done in accordance with common industry practices.

Problems 25 and 26 are based on the following information and illustration.

A flow net is drawn for the homogeneous, isotropic soil beneath an impermeable concrete dam as shown. Beneath the soil lies impermeable bedrock. The upstream water level, H_1, is 3 m above the top of the soil, and the downstream water level, H_2, is 1 m above the top of the soil. The coefficient of permeability, k, is 3×10^{-2} cm/s.

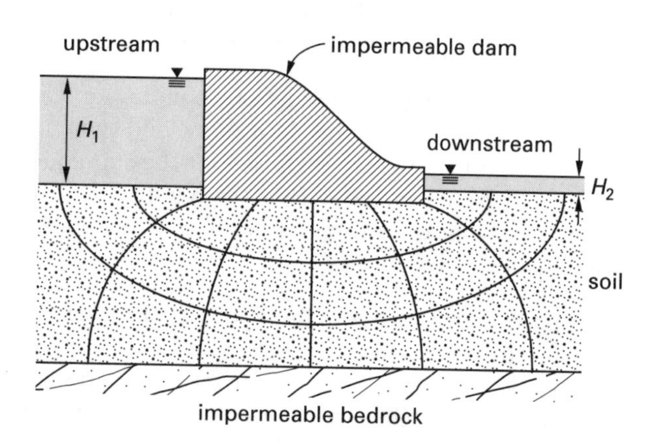

25. The rate of flow per lineal meter of dam width is most nearly

(A) 3×10^{-4} m^2/s
(B) 5×10^{-3} m^2/s
(C) 2×10^{-2} m^2/s
(D) 2×10^{-1} m^2/s

26. If the dam is 10 m wide, the total flow rate of water passing underneath the entire dam is most nearly

(A) 3×10^{-3} m^3/s
(B) 5×10^{-2} m^3/s
(C) 2×10^{-1} m^3/s
(D) 2 m^3/s

Problems 27–29 are based on the following information and illustration. The given table relates the average one-dimensional consolidation of a uniform clay layer to its corresponding time factor.

U_{avg} (%)	T
0	0.000
10	0.008
20	0.031
30	0.071
40	0.126
50	0.197

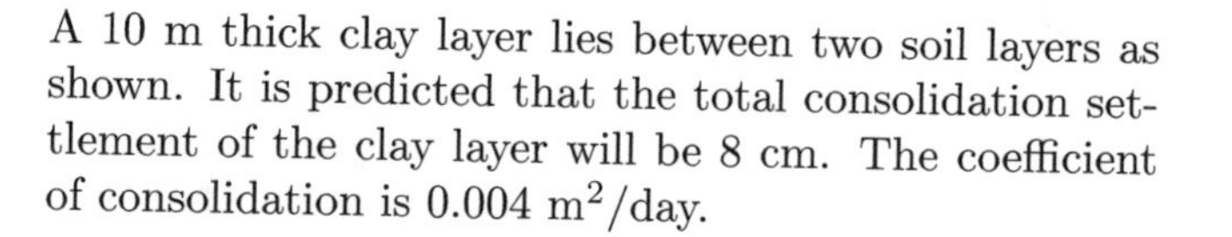
A 10 m thick clay layer lies between two soil layers as shown. It is predicted that the total consolidation settlement of the clay layer will be 8 cm. The coefficient of consolidation is 0.004 m^2/day.

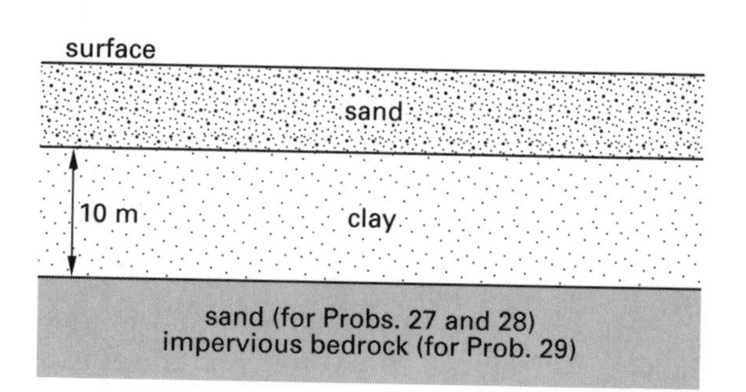

27. Assuming that the clay layer lies between two sand layers, the amount of time required for 20% of the total settlement to occur is most nearly

(A) 39 days
(B) 78 days
(C) 190 days
(D) 780 days

28. Assuming that the clay layer lies between two sand layers, the total amount of settlement after one year is most nearly

(A) 1.0 cm
(B) 2.1 cm
(C) 3.2 cm
(D) 4.3 cm

29. Assuming that the clay layer is bounded by a sand layer on top and impervious bedrock on bottom, the total amount of settlement to occur after one year is most nearly

(A) 1.0 cm
(B) 2.1 cm
(C) 3.2 cm
(D) 4.3 cm

30. The base of a 2 m wide continuous footing is 1 m below the ground surface. The soil under the footing has the following parameters.

$$\rho = 1835 \text{ kg/m}^3$$
$$N_c = 9.6$$
$$\phi = 10°$$
$$N_q = 2.7$$
$$c = 0.0$$
$$N_\gamma = 1.2$$

If a factor of safety of three is required, the allowable bearing capacity of the soil under the footing is most nearly

(A) 17 kPa
(B) 23 kPa
(C) 49 kPa
(D) 70 kPa

Problems 31 and 32 are based on the following information and illustration.

A beam is loaded as shown.

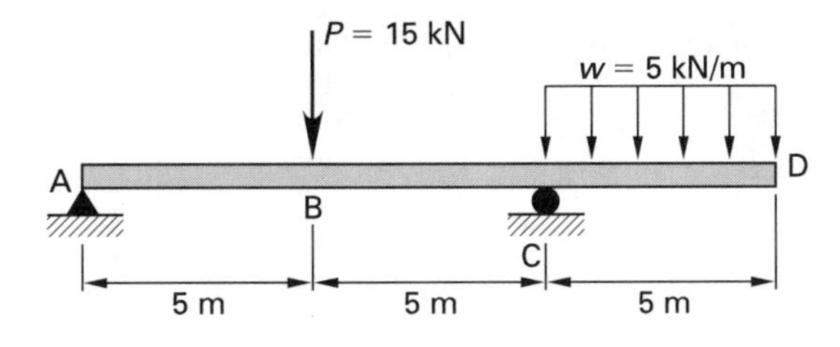

31. The magnitude of the maximum bending moment in the beam is most nearly

(A) 6.3 kN·m
(B) 14 kN·m
(C) 25 kN·m
(D) 63 kN·m

32. If the beam is made entirely of steel and the whole beam has a moment of inertia about the axis of bending of 2.0×10^8 mm^4, the magnitude of the vertical deflection at point D is most nearly

(A) 0.2 mm
(B) 2.3 mm
(C) 23 mm
(D) 50 mm

Problems 33 and 34 are based on the following information and illustration.

A truck is facing in its intended direction of travel along the beam as shown.

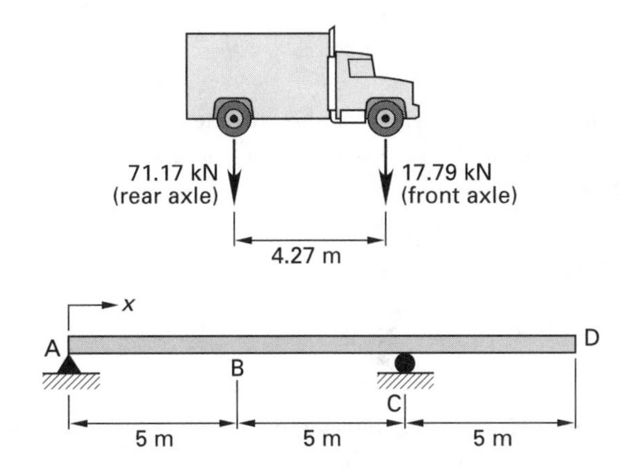

33. For a truck traveling in the direction shown, the magnitude of maximum vertical live load shear at support C is most nearly

(A) 27 kN
(B) 76 kN
(C) 103 kN
(D) 107 kN

34. For a truck traveling in the direction shown, the magnitude of maximum live load bending moment at support C is most nearly

(A) 80 kN·m
(B) 90 kN·m
(C) 140 kN·m
(D) 360 kN·m

Problems 35 and 36 are based on the following information and illustration.

A plane truss span is shown. The roadway behaves as simply supported beam spans between the supporting lower chord truss joints. By convention, positive forces are tensile forces, and negative forces are compressive forces.

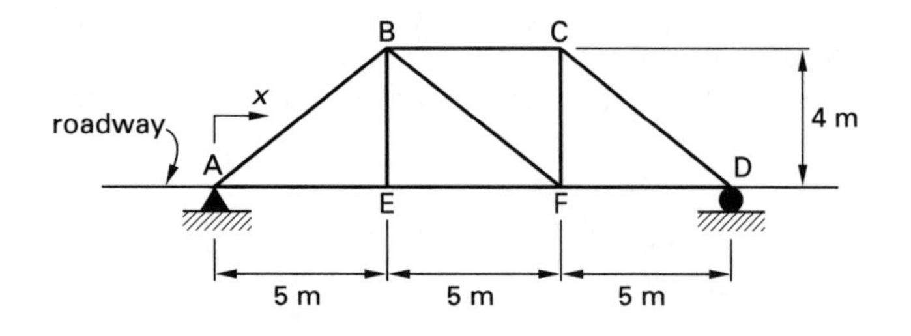

35. In the x-direction as shown, the maximum influence line ordinate for tensile force in member BF is most nearly

(A) 0.18 kN/kN
(B) 0.36 kN/kN
(C) 0.53 kN/kN
(D) 0.71 kN/kN

36. In the x-direction as shown, the maximum influence line ordinate for compressive force in member BF is most nearly

(A) −0.71 kN/kN
(B) −0.53 kN/kN
(C) −0.36 kN/kN
(D) −0.18 kN/kN

Problems 37 and 38 are based on the following information and illustration.

The cross sections of two short concentrically loaded reinforced concrete columns are shown.

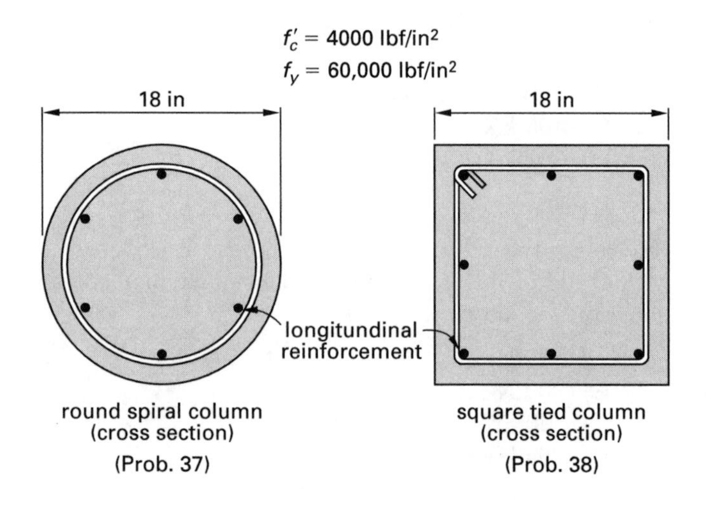

37. For the round spiral column, the applied axial dead load, P_{dead}, is 150 kips, and the applied axial live load, P_{live}, is 350 kips. Based on American Concrete Institute (ACI) strength design and assuming that the longitudinal reinforcing bars are all the same size, the minimum required size of each longitudinal reinforcing bar is

(A) No. 7
(B) No. 8
(C) No. 9
(D) No. 10

38. For the square tied column, the applied axial dead load, P_{dead}, is 150 kips, and the applied axial live load, P_{live}, is 250 kips. Based on ACI strength design and assuming that the longitudinal reinforcing bars are all the same size, the minimum required size of each longitudinal reinforcing bar is

(A) No. 3
(B) No. 4
(C) No. 5
(D) No. 6

Problems 39 and 40 are based on the following information and illustration.

A steel compression member has a fixed support at one end and a pinned support at the other as shown. The total applied design load, P_{total}, consists of a dead load, P_{dead}, of 7 kips and an unspecified live load, P_{live}.

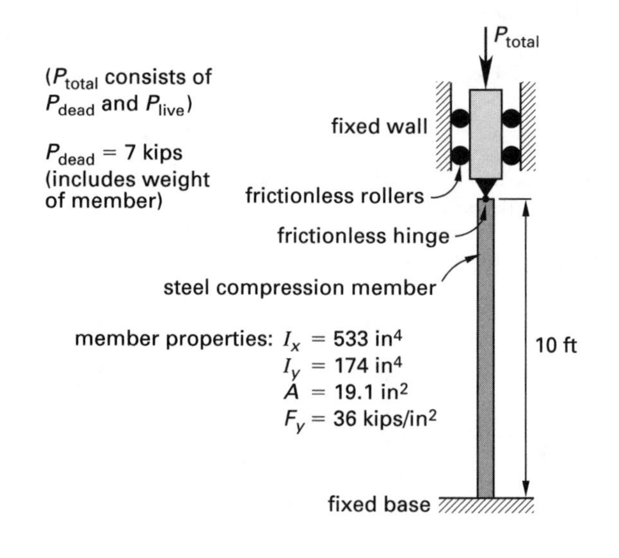

39. In accordance with American Institute of Steel Construction (AISC) Allowable Stress Design (ASD) specifications, the maximum allowable design live load, P_{live}, is most nearly

(A) 6 kips
(B) 370 kips
(C) 390 kips
(D) 2900 kips

40. In accordance with American Institute of Steel Construction (AISC) Load and Resistance Factor Design (LRFD) specifications, the maximum allowable design live load, P_{live}, is most nearly

(A) 340 kips
(B) 400 kips
(C) 550 kips
(D) 650 kips

Problems 41 and 42 are based on the following information and illustration.

A bolted steel tension member is shown. The total applied design load, P_{total}, consists of a dead load, P_{dead}, of 15 kips and an unspecified live load, P_{live}.

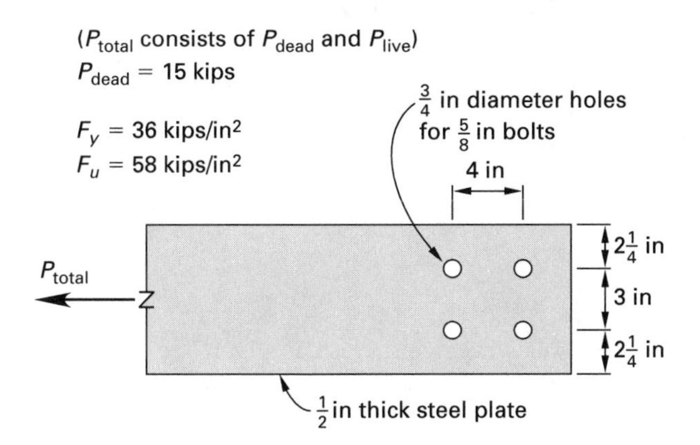

41. In accordance with American Institute of Steel Construction (AISC) Allowable Stress Design (ASD) specifications, the maximum allowable design live load, P_{live}, is most nearly

(A) 50 kips
(B) 66 kips
(C) 80 kips
(D) 87 kips

42. In accordance with American Institute of Steel Construction (AISC) Load and Resistance Factor Design (LRFD) specifications, the maximum allowable design live load, P_{live}, is most nearly

(A) 50 kips
(B) 56 kips
(C) 65 kips
(D) 70 kips

43. A boundary and traverse line are shown.

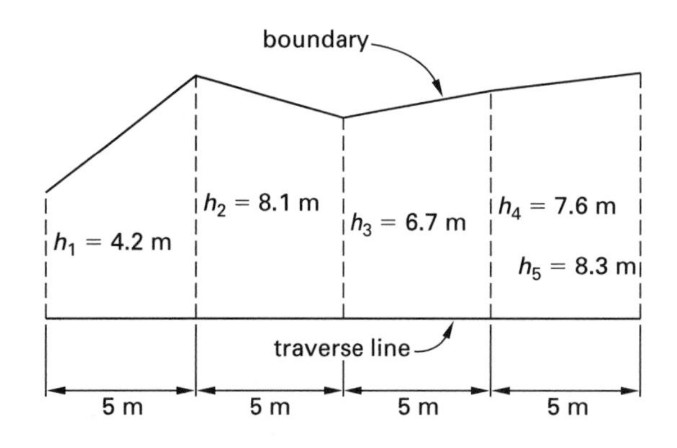

Using Simpson's 1/3 rule, the total area between the boundary and traverse line is most nearly

(A) 141 m²
(B) 143 m²
(C) 148 m²
(D) 151 m²

Problems 44 and 45 are based on the following information.

A back tangent with a +7% grade meets a forward tangent with a −5% grade on a vertical alignment. A 350 m (10-station) horizontal length of vertical curve is placed such that the point of vertical curvature (PVC) is at sta 10+35 at an elevation of 60.0 m.

44. The vertical curve elevation at sta 11+35 is most nearly

(A) 65 m
(B) 67 m
(C) 69 m
(D) 71 m

45. The tangent elevation at the point of vertical intersection (PVI) is most nearly

(A) 66 m
(B) 68 m
(C) 70 m
(D) 72 m

46. A reading of 3.50 m is taken on a 4 m leveling rod that is 0.50 m out of plumb at the top of the rod. The correct reading, when the rod is truly vertical, is most nearly

(A) 3.06 m
(B) 3.47 m
(C) 3.53 m
(D) 3.94 m

Problems 47 and 48 are based on the following information and illustration.

A horizontal curve is laid out with the point of curve (PC) station and the length of long chord (LC) as shown.

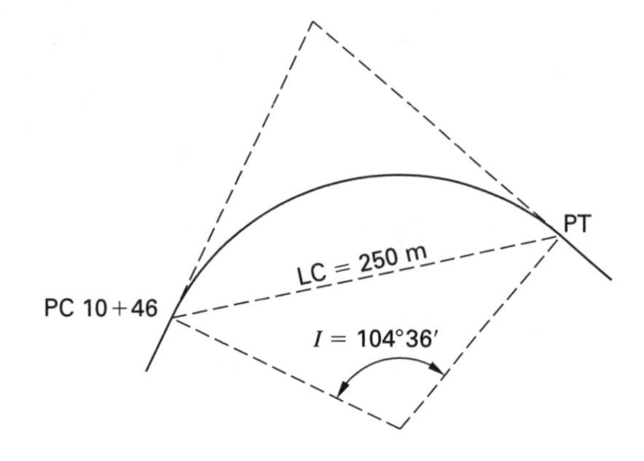

47. The radius of the curve is most nearly

(A) 158 m
(B) 160 m
(C) 316 m
(D) 320 m

48. The point of tangent (PT) is located at station

(A) 13+27.1
(B) 13+28.5
(C) 13+34.4
(D) 13+39.2

49. A one-lane rural road has a 10° curve extending for 700 ft along its centerline. The road is 15 ft wide with 9 ft wide shoulders. The design speed for this road is 45 mi/hr.

The minimum required length of spiral transition onto this road is most nearly

(A) 28 ft
(B) 36 ft
(C) 44 ft
(D) 50 ft

50. The design requirements for a section of highway with a 1.5% grade are as follows.

design speed = 80 km/h
coefficient of friction = 0.35
driver reaction time = 2.0 s
driver eye height = 1.2 m
object (to be avoided) height = 0.2 m

The design braking distance for this highway is most nearly

(A) 45 m
(B) 75 m
(C) 100 m
(D) 120 m

51. A crest on a section of highway consists of a vertical curve with a 1500 m radius and a positive 1% grade followed by a negative 3% grade. The design requirements are as follows.

design speed = 80 km/h
driver eye height = 1.2 m
object (to be avoided) height = 0.2 m
stopping sight distance = 300 m

The minimum required length of vertical curve needed to satisfy the design stopping sight distance is most nearly

(A) 680 m
(B) 700 m
(C) 760 m
(D) 840 m

52. The following wind data is given for a proposed airport site

wind direction	percentage of winds		
	4–15 mph	15–31 mph	31–47 mph
N	5	15	4
NE	1	0	0
E	13	5	3
SE	7	0	0
S	10	1	2
SW	11	0	0
W	19	2	1
NW	1	0	0

The best main runway orientation for heavy aircraft (that is, aircraft greater than 12,500 lbf in weight) is most nearly

(A) N-S
(B) NE-SW
(C) E-W
(D) SE-NW

53. The design AASHTO structural number for a road is 4. The material specifications are as follows.

material	layer thickness	experience coefficient
sandy gravel subbase	10 in	0.11
crushed stone base course	6 in	0.14

If a high-stability plant mix asphalt concrete surface course with an experience coefficient of 0.44 is to be placed on top of the specified subbase and base course materials, the required surface course thickness is most nearly

(A) 3 in
(B) 4 in
(C) 5 in
(D) 6 in

54. A road leading to a stone quarry is traveled by 40 trucks, with each truck making an average of 10 trips per day. When fully loaded, each truck consists of a front single axle transmitting a force of 10,000 lbf and two rear tandem axles, each axle transmitting a force of 20,000 lbf. The load equivalency factor for the front 10,000 lbf single axle is 0.0877. The load equivalency factor for each rear tandem 20,000 lbf axle is 0.1206.

The 18,000 lbf equivalent single axle load (ESAL) for the truck traffic on this road for five years is most nearly

(A) 0.33
(B) 130
(C) 48,000
(D) 240,000

55. A wastewater sample contains 350 mg/L of suspended solids. Primary sedimentation facilities remove 65% of the suspended solids. The sludge produced from primary sedimentation processes contains 5% solids. The amount of sludge produced per liter of wastewater is most nearly

(A) 1.8 g
(B) 2.3 g
(C) 3.5 g
(D) 4.6 g

56. Testing on a water sample shows the following analysis.

cation	cation mg/L	anion	anion mg/L
Na^+	23	Cl^-	35
Ca^{2+}	10	SO_4^{2-}	16
Mg^{2+}	15	NO_3^-	9

The hardness of the water is most nearly

(A) 25 mg/L as $CaCO_3$ equivalents
(B) 62 mg/L as $CaCO_3$ equivalents
(C) 87 mg/L as $CaCO_3$ equivalents
(D) 1300 mg/L as $CaCO_3$ equivalents

57. A rectangular channel is 1.5 m wide, 3 m deep, and 25 m long. The design flow rate of the wastewater in the channel is 0.5 m^3/s. The approach velocity in the channel is most nearly

(A) 0.1 m/s
(B) 0.2 m/s
(C) 0.3 m/s
(D) 0.5 m/s

58. A water sample has a total hardness of 75 mg/L as $CaCO_3$ equivalents. The hardness is caused solely by magnesium and is carbonate hardness. If the water is to be softened to 5 mg/L as $CaCO_3$ equivalents by the addition of calcium hydroxide, the required calcium hydroxide concentration is most nearly

(A) 0.0005 M
(B) 0.0007 M
(C) 0.0014 M
(D) 0.0075 M

Problems 59 and 60 are based on the following information and illustration.

A primary clarification system has an average wastewater flow of 7000 m^3/day with a maximum flow of 16 000 m^3/day and a minimum flow of 4000 m^3/day. Performance of the system is given by the curve shown. The system uses two side-by-side rectangular basins and is designed for an estimated BOD removal of 35% at peak flow. The minimum hydraulic residence time is one hour.

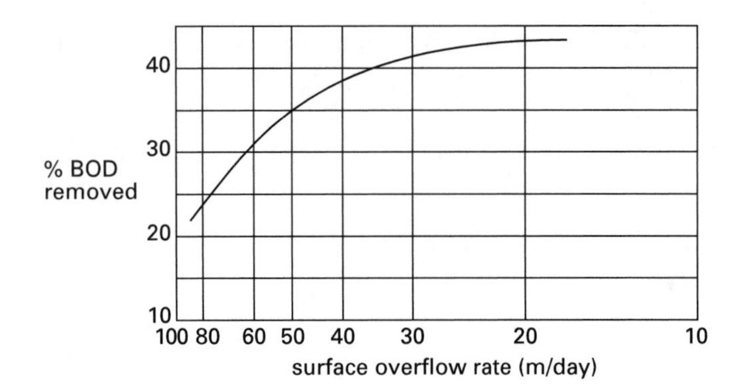

59. Assuming that both basins are always operational, the minimum required basin depth per unit length is most nearly

(A) 0.9 m
(B) 1.0 m
(C) 1.8 m
(D) 2.1 m

60. If the length of each basin is 20 m, the width of each basin is 5 m, and the depth of each basin is 2.5 m, the hydraulic loading rate is most nearly

(A) 1.4 $m^3/m^2{\cdot}h$
(B) 1.6 $m^3/m^2{\cdot}h$
(C) 2.9 $m^3/m^2{\cdot}h$
(D) 3.3 $m^3/m^2{\cdot}h$

SOLUTIONS FOR THE PRACTICE EXAM

1. Simpson's 1/3 rule can be adapted for numerical analysis as shown by

$$\text{area} = \frac{w\left[h_1 + 2\sum h_{\text{odds}} + 4\sum h_{\text{evens}} + h_n\right]}{3}$$

$$[n = \text{odd number}]$$

$$\int_{x_1}^{x_n} f(x)dx = \left(\frac{\Delta x}{3}\right)\left(\begin{array}{l} y_1 + 2\sum\limits_{i=3,5,7,\ldots}^{i=n-2} y_i \\ +4\sum\limits_{i=2,4,6,\ldots}^{i=n-1} y_i \\ +y_n \end{array}\right)$$

The numerical values are

$$\begin{aligned} n &= \frac{\text{upper limit} - \text{lower limit}}{\Delta x} = \frac{3\ \text{km} - 0\ \text{km}}{0.5\ \text{km}} + 1 \\ &= 7 \\ x_1 &= 0\ \text{km} \\ x_2 &= 0.5\ \text{km} \\ x_3 &= 1\ \text{km} \\ x_4 &= 1.5\ \text{km} \\ x_5 &= 2\ \text{km} \\ x_6 &= 2.5\ \text{km} \\ x_7 &= 3\ \text{km} \end{aligned}$$

y_i is the numerical value of the function $y = f(x)$ evaluated at $x = x_i$.

Substituting appropriate values gives

$$\begin{aligned} &\int_{x_1=0}^{x_7=3} \left(\frac{e^x}{(e+x)^x}\right) dx \\ &= \left(\frac{\Delta x}{3}\right)\left(y_1 + 2\sum_{i=3,5}^{i=n} y_i + 4\sum_{i=2,4,6}^{i=n-1} y_i + y_7\right) \\ &= \left(\frac{0.5\ \text{km}}{3}\right)\left[\begin{array}{l} 1\ \text{km} + (2)(0.7311\ \text{km} + 0.3319\ \text{km}) \\ +(4)\left(\begin{array}{l} 0.9190\ \text{km} + 0.5173\ \text{km} \\ +0.1958\ \text{km} \end{array}\right) \\ +0.1074\ \text{km} \end{array}\right] \\ &= 1.627\ \text{km}^2 \quad (1.63\ \text{km}^2) \end{aligned}$$

Answer is B.

Solutions 2 and 3 are based on the following information.

The number of degrees of freedom is the number of displacements each node can undergo. For beam element AB, each node (A and B) can undergo translation in the x- and y-directions, and rotations about the x-, y-, and z-axes.

2. The number of degrees of freedom is

$$\begin{aligned} &(2\ \text{nodes})\left(5\ \frac{\text{degrees of freedom}}{\text{node}}\right) \\ &\quad = 10\ \text{degrees of freedom} \end{aligned}$$

Answer is C.

3. The number of degrees of freedom is

$$\begin{aligned} &(3\ \text{nodes})\left(5\ \frac{\text{degrees of freedom}}{\text{node}}\right) \\ &\quad = 15\ \text{degrees of freedom} \end{aligned}$$

Answer is D.

4. The true error with respect to the final answer is

$$\begin{aligned} E_t &= \text{true value} - \text{approximation} = 3.142 - 3.132 \\ &= 0.010 \end{aligned}$$

Answer is A.

5. The true percent relative error with respect to the final solution is

$$\begin{aligned} \varepsilon_t &= \left(\frac{\text{true value} - \text{approximation}}{\text{true value}}\right)(100\%) \\ &= \left(\frac{3.142 - 3.132}{3.142}\right)(100\%) \\ &= 0.3183\% \quad (0.3\%) \end{aligned}$$

Answer is C.

6. The approximate percent relative error for the two iterations is

$$\begin{aligned} \varepsilon_t &= \left(\frac{\begin{array}{l}\text{present approximation} \\ \quad - \text{previous approximation}\end{array}}{\text{present approximation}}\right)(100\%) \\ &= \left(\frac{3.132 - 3.110}{3.132}\right)(100\%) \\ &= 0.7024\% \quad (0.7\%) \end{aligned}$$

Answer is D.

Solutions 7–9 are based on the following information and illustration.

Problems 7–9 can be solved with critical path method (CPM) calculations. Since the project is to start on January 1, it is easy to designate the actual start time as the end of the previous day, December 31, and designate it as day 0 with January 1 being designated as day 1.

Determination of the earliest start time (EST) and earliest finish time (EFT) for an activity is done by a forward pass through the diagram.

The EST of an activity is calculated as the maximum of the EFTs of the activities preceding it. For example, activity A has no activities preceding it, so it has an EST of 0. The EFT of this activity is given by

$$\text{EFT of activity A} = \text{EST} + \text{duration} = 0 + 8 = 8$$

Similarly, activity B has an EST of 0 and an EFT of 5.

Activity D is preceded only by activity B, so its EST is the EFT of activity B. That is, the EST of activity D is 5 and the EFT is 14.

Activity F is preceded by two activities, A and D, so the EST of activity F is the maximum EFT of the two. That is, the EST of activity F is 14 and its EFT is 17.

Similar calculations show that the minimum project duration is 17 days (EFT for activity F is 17).

Determination of the latest start time (LST) and latest finish time (LFT) for an activity is done by a backward pass through the diagram using the minimum project duration as the starting point. That is, 17 is the starting point for these calculations.

The LFT of an activity is the minimum LST of the activities following it. For example, the LFTs of the activities preceding the project finish are all 17 since no activities follow them. Accordingly, the LFTs of activities F and G are each 17. The LST of activity F is given by

$$\text{LST of activity F} = \text{LFT} - \text{duration} = 17 - 3 = 14$$

Similarly, the LST of activity G is $17 - 5 = 12$.

Activity D is preceded only by activity F, so its LFT is the LST of activity F. That is, the LFT of activity D is 14 and the LST of activity D is 5.

Activity B is preceded by two activities, D and G, so its LFT is the minimum LST of the two. That is, the LFT of activity B is 5.

The total float time (TF) of an activity is determined either by subtracting the EFT from the LFT or by subtracting the EST from the LST. For example, the TF of activity G is

$$\text{TF of activity G} = \text{LFT} - \text{EFT} = 17 - 10 = 7$$

Making a summary table is a good way to organize these results.

activity	duration	EST	EFT	LST	LFT	TF
A	8	0	8	6	14	6
B	5	0	5	0	5	0
C	7	0	7	9	16	9
D	9	5	14	5	14	0
E	3	5	8	13	16	8
F	3	14	17	14	17	0
G	5	5	10	12	17	7
H	1	8	9	16	17	8

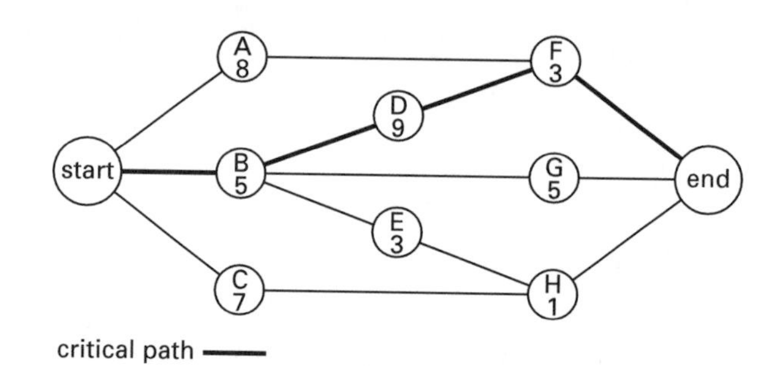

7. The earliest date this project can be completed is the maximum EFT of the entire project, and the EFT of activity F is 17. Therefore, using January 1 as day 1, day 17 falls on January 17.

Answer is D.

8. The total float for activity A can be seen from the summary table as being six days.

Answer is C.

9. The LST for activity E is 13. Therefore, using January 1 as day 1, the latest day that activity E can start is January 13.

Answer is B.

10. Each lift is 0.1 m thick. The mass of each square meter of lift is

$$(0.1 \text{ m})\left(1630 \ \frac{\text{kg}}{\text{m}^3}\right) = 163 \text{ kg/m}^2$$

The amount of bentonite in the soil is to be 5% dry basis. Therefore, the amount of bentonite in each lift is

$$(0.05)\left(163\ \frac{\text{kg}}{\text{m}^2}\right) = 8.15\ \text{kg/m}^2 \quad (8.2\ \text{kg/m}^2)$$

Answer is A.

11. **Answer is B.**

12. **Answer is A.**

13. Substituting the given seven-day values into the rearranged equation for BOD exertion gives

$$y_t = L(1 - e^{-kt})$$

$$L = \frac{y_t}{1 - e^{-kt}} = \frac{211\ \frac{\text{mg}}{\text{L}}}{1 - e^{-\left(0.14\ \frac{1}{\text{days}}\right)(7\ \text{days})}}$$

$$= 338\ \text{mg/L} \quad (340\ \text{mg/L})$$

Answer is C.

14. From Prob. 13, $L = 338$ mg/L. Substituting appropriate values into the equation for BOD exertion gives

$$y_t = L(1 - e^{-kt})$$

$$y_5 = \left(338\ \frac{\text{mg}}{\text{L}}\right)\left[1 - e^{-\left(0.14\ \frac{1}{\text{days}}\right)(5\ \text{days})}\right]$$

$$= 170\ \text{mg/L}$$

Answer is B.

15. The dissolved oxygen content (DO) at 760 mm Hg can be found by linear interpolation from the table given in the problem statement.

$$\frac{\text{DO}_{23.3^\circ\text{C}} - 8.7\ \frac{\text{mg}}{\text{L}}}{8.5\ \frac{\text{mg}}{\text{L}} - 8.7\ \frac{\text{mg}}{\text{L}}} = \frac{23.3^\circ\text{C} - 23^\circ\text{C}}{24^\circ\text{C} - 23^\circ\text{C}}$$

$$\text{DO}_{23.3^\circ\text{C}} = \frac{\left(8.5\ \frac{\text{mg}}{\text{L}} - 8.7\ \frac{\text{mg}}{\text{L}}\right)(23.3^\circ\text{C} - 23^\circ\text{C})}{24^\circ\text{C} - 23^\circ\text{C}} + 8.7\ \frac{\text{mg}}{\text{L}}$$

$$= 8.64\ \text{mg/L}$$

Oxygen is only slightly soluble in water and does not react with water chemically. Therefore, Henry's law is applicable, and oxygen's solubility is directly proportional to its partial pressure. Therefore, the percent saturation of oxygen in the given fresh water sample at an atmospheric pressure of 730 mm Hg is

$$\%\ \text{saturation} = \left(\frac{730\ \text{mm Hg}}{760\ \text{mm Hg}}\right)\left(\frac{5.7\ \frac{\text{mg}}{\text{L}}}{8.64\ \frac{\text{mg}}{\text{L}}}\right)(100\%)$$

$$= 63.4\%$$

Answer is A.

16. With a turbine at the pipe outlet, set $\text{v}_2 = 0$ to get the head supplied to the turbine. The head supplied to the turbine can be calculated from the energy equation.

$$\frac{p_1}{\gamma} + z_1 + \frac{\text{v}_1^2}{2g} = \frac{p_2}{\gamma} + z_2 + \frac{\text{v}_2^2}{2g} + h_f + h_{\text{turbine}}$$

$$h_{\text{turbine}} = \frac{p_1}{\gamma} + z_1 + \frac{\text{v}_1^2}{2g} - \frac{p_2}{\gamma} - z_2 - \frac{\text{v}_2^2}{2g} - h_f$$

$$= 0 + 200\ \text{m} + 0 - 0 - 180\ \text{m} - 18\ \text{m}$$

$$= 2\ \text{m}$$

The power output of the turbine is

$$\dot{W} = Q\gamma h_{\text{turbine}}\eta = Q\rho g h_{\text{turbine}}\eta$$

$$= \left(4.92\ \frac{\text{m}^3}{\text{s}}\right)\left(1000\ \frac{\text{kg}}{\text{m}^3}\right)\left(9.81\ \frac{\text{m}}{\text{s}^2}\right)(2\ \text{m})(0.85)$$

$$= 82\,051\ \text{W} \quad (82\ \text{kW})$$

(Note that a turbine is similar to a pump except that it is energized by fluid instead of the fluid being energized by it. Therefore, the ideal power is *multiplied* by the efficiency, η, not divided by it, as is done in the power equation for pumps.)

Answer is A.

17. The Hydraulic Elements Graph for Circular Sewers in the "Civil Engineering" section of the NCEES Handbook can be used to avoid having to calculate the hydraulic radius and cross-sectional area of flow for partial flow in circular pipes.

The full-flow variables are first determined to be

$$Q_f = 15\ \text{m}^3/\text{s}$$

$$\text{v}_f = \frac{Q_f}{A_f} = \frac{15\ \frac{\text{m}^3}{\text{s}}}{\frac{\pi(1.5\ \text{m})^2}{4}} = 8.49\ \text{m/s}$$

The ordinate to be located on the hydraulic elements graph is

$$\frac{d}{D} = \frac{0.50\ \text{m}}{1.5\ \text{m}} = 0.333$$

From the relevant curves on the Hydraulic Elements Graph for Circular Sewers in the NCEES Handbook, the Q/Q_f and v/v_f ratios are

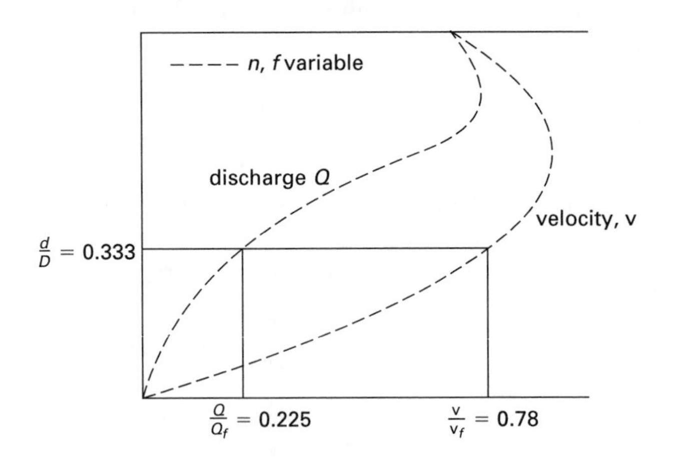

The flow rate with a depth of flow of 0.50 m is

$$\frac{Q}{Q_f} = 0.225$$

$$Q = 0.225Q_f = (0.225)\left(15\ \frac{\text{m}^3}{\text{s}}\right) = 3.38\ \text{m}^3/\text{s} \quad (3.4\ \text{m}^3/\text{s})$$

Answer is B.

18. From Prob. 17, $\text{v}/\text{v}_f = 0.78$. Therefore, the flow velocity when the depth of flow is 0.50 m is

$$\frac{\text{v}}{\text{v}_f} = 0.78$$

$$\text{v} = 0.78\text{v}_f = (0.78)\left(8.49\ \frac{\text{m}}{\text{s}}\right) = 6.62\ \text{m/s} \quad (6.6\ \text{m/s})$$

Answer is D.

19. The overall runoff coefficient is determined by proportioning each individual runoff coefficient in accordance with the relative area it covers.

$$\begin{aligned}\text{total area} &= 7500\ \text{m}^2\ \text{lawn} + 2000\ \text{m}^2\ \text{gravel road} \\ &\quad + 500\ \text{m}^2\ \text{roof surface} \\ &= 10\,000\ \text{m}^2\end{aligned}$$

Therefore, by weighting each runoff coefficient with the relative area it covers, the overall runoff coefficient is

$$\begin{aligned}C_{\text{overall}} &= \left(\frac{7500\ \text{m}^2}{10\,000\ \text{m}^2}\right)(0.20) + \left(\frac{2000\ \text{m}^2}{10\,000\ \text{m}^2}\right)(0.15) \\ &\quad + \left(\frac{500\ \text{m}^2}{10\,000\ \text{m}^2}\right)(0.80) \\ &= 0.22\end{aligned}$$

Answer is B.

20. From the Sewage Flow Ratio Curves in the "Civil Engineering" section of the NCEES Handbook, the minimum flow ratio is conservatively determined from curve E_2, unless otherwise specified, to be about 0.32 for a population of 20,000. Therefore, the estimated minimum sewage flow is

$$\frac{Q_{\text{min}}}{Q_{\text{avg}}} = 0.32$$

$$Q_{\text{min}} = 0.32Q_{\text{avg}} = (0.32)\left(8000\ \frac{\text{m}^3}{\text{day}}\right) = 2560\ \text{m}^3/\text{day} \quad (2600\ \text{m}^3/\text{day})$$

Answer is B.

21. From the Sewage Flow Ratio Curves in the "Civil Engineering" section of the NCEES Handbook, the maximum flow ratio is conservatively determined from curve C, unless otherwise specified, to be about 3.4 for a population of 20,000. Therefore, the estimated maximum sewage flow is

$$\frac{Q_{\text{max}}}{Q_{\text{avg}}} = 3.4$$

$$Q_{\text{max}} = 3.4Q_{\text{avg}} = (3.4)\left(8000\ \frac{\text{m}^3}{\text{day}}\right) = 27\,200\ \text{m}^3/\text{day} \quad (27\,000\ \text{m}^3/\text{day})$$

Answer is C.

22. Professional engineers have primary obligations to society, clients, and to the profession. Engineers have secondary obligations to themselves.

Answer is D.

23. The "engineer of record" refers to the designer.

Answer is A.

24. Even if it is common industry practice to design for only uniform roof live loads, prevailing building codes set the standards that must be followed. Therefore, unbalanced roof live loads should have been taken into consideration in the structural design.

Answer is A.

25. For the flow net shown, $N_f = 3$ and $N_d = 6$. The flow per lineal meter of dam width is given by

$$Q = kH\left(\frac{N_f}{N_d}\right) = k(H_1 - H_2)\left(\frac{N_f}{N_d}\right)$$
$$= \left[\left(3 \times 10^{-2}\ \frac{\text{cm}}{\text{s}}\right)\left(\frac{1\ \text{m}}{100\ \text{cm}}\right)\right](3\ \text{m} - 1\ \text{m})\left(\frac{3}{6}\right)$$
$$= 3 \times 10^{-4}\ \text{m}^2/\text{s}$$

Answer is A.

26. From Prob. 25, $Q = 3 \times 10^{-4}\ \text{m}^2/\text{s}$ per lineal meter of dam width. For a 10 m wide dam, let $w = 10$ m. The total flow rate under the dam is

$$Q_{\text{total}} = Qw = \left(3 \times 10^{-4}\ \frac{\text{m}^2}{\text{s}}\right)(10\ \text{m})$$
$$= 3 \times 10^{-3}\ \text{m}^3/\text{s}$$

Answer is A.

27. For 20% of the total settlement, $T = 0.031$. With sand layers bounding the clay layer, $H = 10\ \text{m}/2 = 5$ m for the doubly drained clay layer. Therefore, the amount of time required for 20% settlement is

$$t_{20} = \frac{T_{20}H^2}{C_v} = \frac{(0.031)(5\ \text{m})^2}{0.004\ \frac{\text{m}^2}{\text{day}}}$$
$$= 193.75\ \text{days}\quad (190\ \text{days})$$

Answer is C.

28. $t = 1$ yr $= 365$ days and $H = 10\ \text{m}/2 = 5$ m for doubly drained clay. The time factor is

$$T = \frac{C_v t}{H^2} = \frac{\left(0.004\ \frac{\text{m}^2}{\text{day}}\right)(365\ \text{days})}{(5\ \text{m})^2} = 0.058$$

Linear interpolation of the time factor table for $T = 0.058$ gives $U_{\text{avg}} = 26.8\%$. Therefore, based on a total settlement of 8 cm, the amount of settlement after one year is

$$S = (26.8\%)\left(\frac{1}{100\%}\right)(8\ \text{cm}) = 2.1\ \text{cm}$$

Answer is B.

29. The impervious bedrock bounding one side of the clay makes this a singly drained clay layer for which $H = 10$ m. The time factor is

$$T = \frac{C_v t}{H^2} = \frac{\left(0.004\ \frac{\text{m}^2}{\text{day}}\right)(365\ \text{days})}{(10\ \text{m})^2} = 0.015$$

Linear interpolation of the time factor table for $T = 0.015$ gives $U_{\text{avg}} = 13.0\%$. Therefore, based on a total settlement of 8 cm, the amount of settlement after one year is approximately

$$S = (0.13)(8\ \text{cm}) = 1.04\ \text{cm}\quad (1.0\ \text{cm})$$

Answer is A.

30. The soil is cohesionless ($c = 0$). Therefore, the ultimate bearing capacity of the soil is given by

$$q_{\text{ultimate}} = cN_c + \gamma D_f N_q + 0.5\gamma B N_\gamma$$
$$= cN_c + \rho g D_f N_q + 0.5\rho g B N_\gamma$$
$$= (0)(9.6) + \left(1835\ \frac{\text{kg}}{\text{m}^3}\right)\left(9.81\ \frac{\text{m}}{\text{s}^2}\right)(1\ \text{m})(2.7)$$
$$+ (0.5)\left(1835\ \frac{\text{kg}}{\text{m}^3}\right)\left(9.81\ \frac{\text{m}}{\text{s}^2}\right)(2\ \text{m})(1.2)$$
$$= 70\,205\ \text{Pa}$$

$$q_{\text{net}} = q_{\text{ultimate}} - \rho g D_f$$
$$= 70\,205\ \text{Pa} - \left(1835\ \frac{\text{kg}}{\text{m}^3}\right)\left(9.81\ \frac{\text{m}}{\text{s}^2}\right)(1\ \text{m})$$
$$= 52\,203\ \text{Pa}\quad (52.2\ \text{kPa})$$

The allowable bearing capacity is

$$q_a = \frac{q_{\text{ultimate}}}{\text{FS}} = \frac{52.2\ \text{kPa}}{3} = 17.4\ \text{kPa}$$

Answer is A.

31. One way to determine the answer to this problem is to construct shear and moment diagrams. The first step is to determine the reactions at A and C. The change in shear is the area under the applied loading, and the change in moment is the area under the shear diagram.

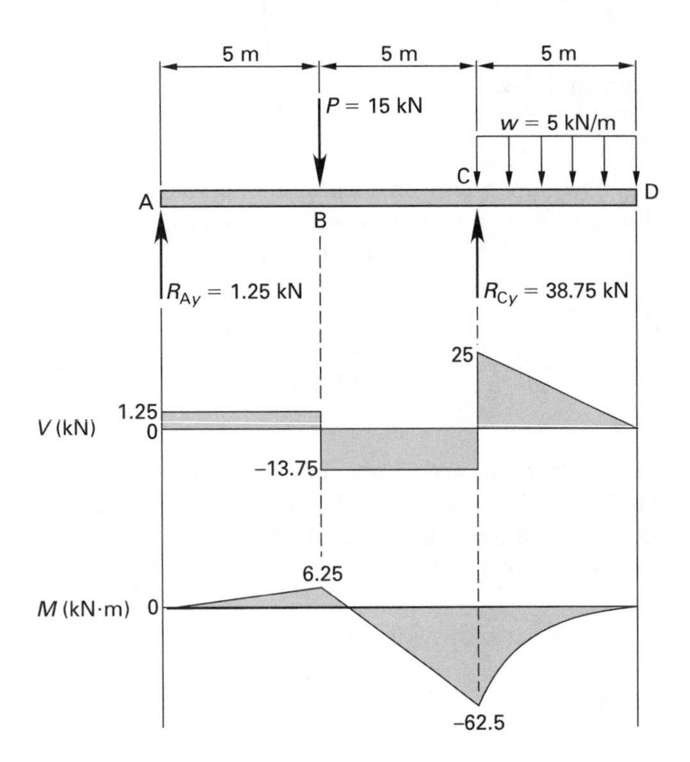

From the moment diagram, the largest magnitude of bending moment is found to be −62.5 kN·m at support C. Therefore, the maximum value of bending moment is 62.5 kN·m.

Answer is D.

32. One way to determine the answer to this problem is by using the principle of virtual work to find the vertical deflection at point D.

$$\Delta_{\mathrm{D}} = \sum\left\{\int m\left(\frac{M}{EI}\right)dx\right\}$$

The moment functions in the directions indicated by the local x-coordinate for each beam segment under the actual loading are shown on the following moment diagram.

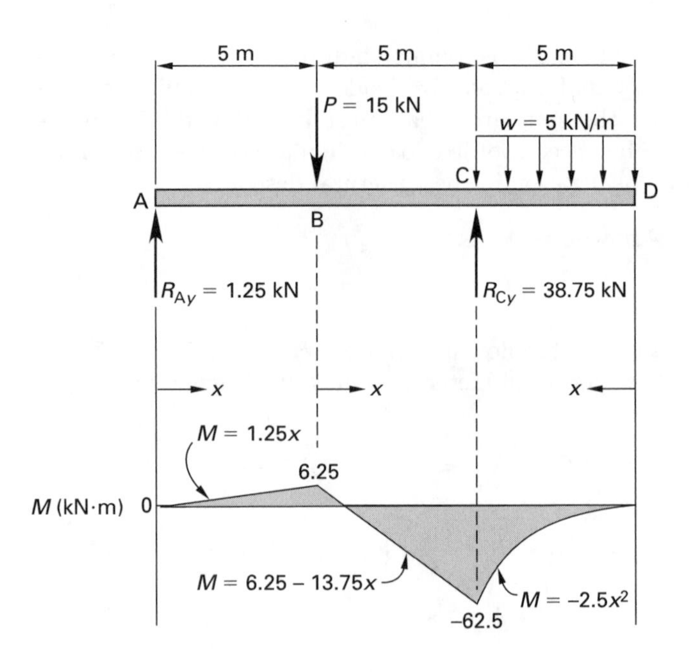

The moment functions in the directions indicated by the local x-coordinate for each beam segment under the virtual unit loading at point D are shown on the following moment diagram.

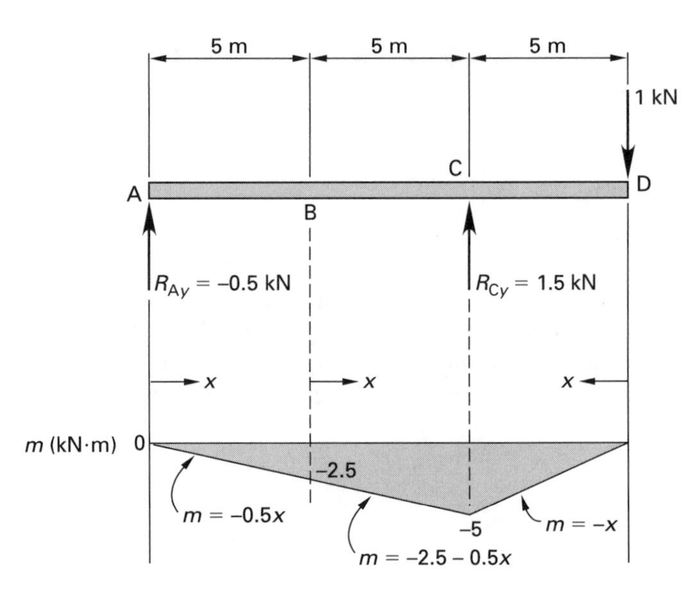

A table can be constructed to summarize the moment functions as follows.

function	region A-B $0 \le x \le 5$	region B-C $0 \le x \le 5$	region C-D $0 \le x \le 5$
M	$1.25x$	$6.25 - 13.75x$	$-2.5x^2$
m	$-0.5x$	$-2.5 - 0.5x$	$-x$
EI	EI	EI	EI

The modulus of elasticity of steel is

$$E = 2.1 \times 10^{11} \text{ Pa}$$

From the principle of virtual work, the vertical deflection at point D can now be calculated to be

$$\begin{aligned}\Delta_{\text{D}} &= \int_0^5 \frac{(-0.5x)(1.25x)}{EI}\,dx \\ &\quad + \int_0^5 \frac{(-2.5 - 0.5x)(6.25 - 13.75x)}{EI}\,dx \\ &\quad + \int_0^5 \frac{(-x)(-2.5x^2)}{EI}\,dx \\ &= \left(\frac{1}{EI}\right)\left(\begin{array}{l}\left[\dfrac{-0.625x^3}{3}\right]_0^5 \\ + \left[-15.625x + \dfrac{31.25x^2}{2} + \dfrac{6.875x^3}{3}\right]_0^5 \\ + \left[\dfrac{2.5x^4}{4}\right]_0^5\end{array}\right) \\ &= \left[\frac{\begin{array}{c}-26.042 \text{ kN}^2\text{·m}^3 \\ + 598.958 \text{ kN}^2\text{·m}^3 \\ + 390.625 \text{ kN}^2\text{·m}^3\end{array}}{\begin{array}{c}(2.1 \times 10^{11} \text{ Pa})\left(\dfrac{1 \text{ kPa}}{1000 \text{ Pa}}\right) \\ \times (2.0 \times 10^8 \text{ mm}^4)\left(\dfrac{1 \text{ m}}{1000 \text{ mm}}\right)^4\end{array}}\right]\left(\frac{1000 \text{ mm}}{1 \text{ m}}\right) \\ &= 22.9 \text{ mm}\end{aligned}$$

Since the unit load in the virtual force system was downward and the answer is positive in sign, the actual deflection is also downward.

Answer is C.

33. This problem can be solved by constructing and using an influence line for vertical shear at support C. The influence line is constructed by plotting the change in response on a free-body diagram of a section of beam at support C as a unit load travels across the structure.

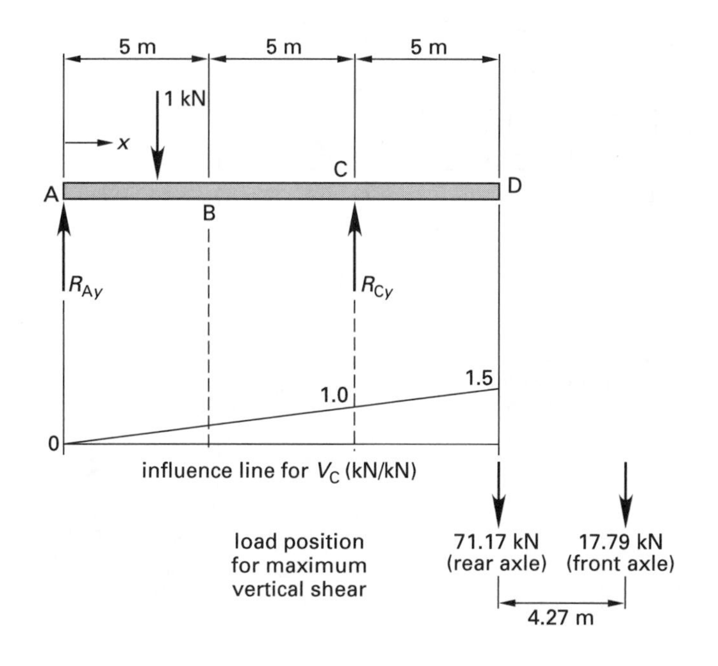

The load position shown gives the maximum response in the beam at support C for the specified direction of travel. The magnitude of maximum vertical shear can be found by superposition to be

$$V_{\text{C,max}} = \left(1.5 \ \frac{\text{kN}}{\text{kN}}\right)(71.17 \text{ kN}) = 106.8 \text{ kN} \quad (107 \text{ kN})$$

(Note that this problem asked for the magnitude of maximum vertical shear at support C for a truck *traveling in the direction as shown.* If this problem had asked for the magnitude of maximum vertical shear at support C *for the given truck axle configuration,* the axle load positions would have to be switched around and both axle loads placed on the influence line with the heavier axle load on the larger influence line ordinate and the lighter axle load on the smaller influence line ordinate. This would result in a larger numerical answer.)

Answer is D.

34. This problem can be solved by constructing and using an influence line for bending moment at support C. The influence line is constructed by plotting the change in response on a free-body diagram of a section of beam at support C as a unit load travels across the structure.

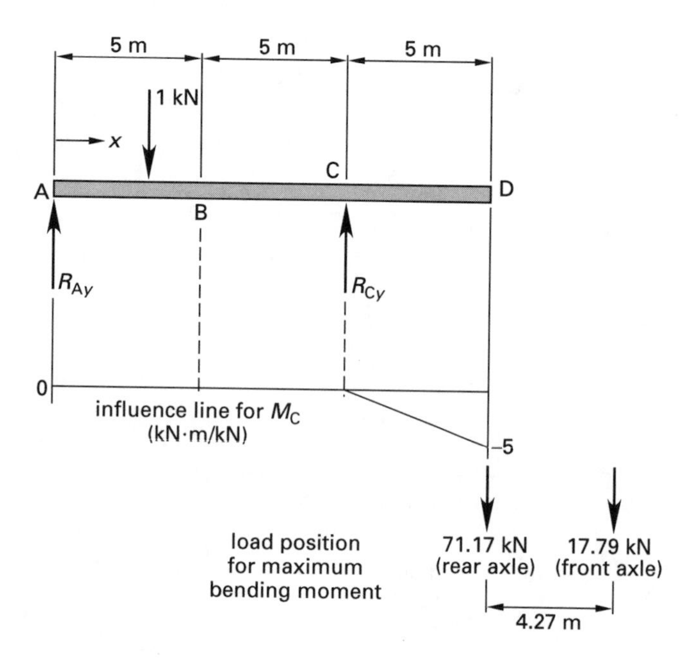

The load position shown gives the maximum response in the beam at support C for the specified direction of travel. The magnitude of maximum bending moment can be found by superposition to be

$$M_{C,\text{max}} = \left|\left(-5\ \frac{\text{kN·m}}{\text{kN}}\right)(71.17\ \text{kN})\right|$$
$$= 355.8\ \text{kN·m}\quad (360\ \text{kN·m})$$

(Note that this problem asked for the magnitude of maximum bending moment at support C for a truck *traveling in the direction as shown.* If this problem had asked for the magnitude of maximum bending moment at support C *for the given truck axle configuration,* the axle load positions would have to be switched around and both axle loads placed on the influence line with the heavier axle load on the larger influence line ordinate and the lighter axle load on the smaller influence line ordinate. This would result in a larger numerical answer.)

Answer is D.

35. This problem can be solved by constructing an influence line for the force in member BF. The influence line is constructed by plotting the change in response in member BF as a unit load travels across the structure.

When moving the unit load across a truss structure, it is important to understand that truss loads can only be applied to truss joints, not between joints. Therefore, when a load is placed between two truss joints, this load must be distributed between the two joints. For this problem, it is stated that the roadway acts as simple beam spans between truss joints. Therefore, when a load is placed between truss joints, the two joints adjoining the beam span act as simple beam supports and the magnitudes of the loads applied to the two truss joints are the same as for the calculated reaction forces of this beam.

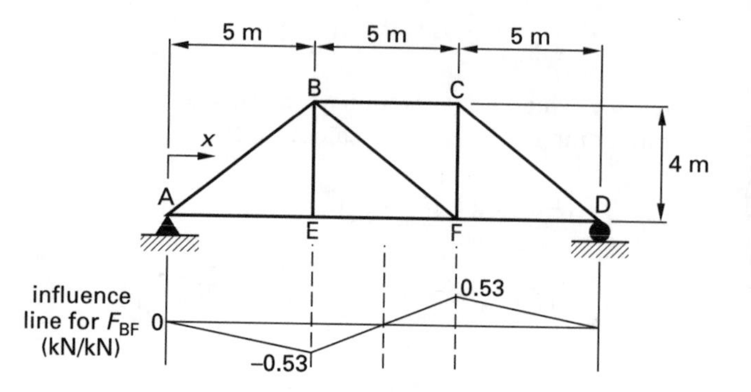

As can be seen from the influence line, the maximum ordinate for tensile force in member BF is 0.53.

Answer is C.

36. As can be seen from the influence line in Prob. 35, the maximum ordinate for compressive force in member BF is found to be −0.53.

Answer is B.

37. Determine the amount of reinforcing steel required by the minimum required reinforcement ratio, ρ_g, of 0.01. Since this ratio is defined as $\rho_g = A_s/A_g$, the minimum area of reinforcing steel required is

$$A_s = \rho_g A_g = (0.01)\left[\frac{\pi(18\ \text{in})^2}{4}\right]$$
$$= 2.54\ \text{in}^2$$

Now determine the required amount of reinforcing steel based on the factored axial load, P_u.

$$P_u = 1.4P_{\text{dead}} + 1.7P_{\text{live}}$$
$$= (1.4)(150\ \text{kips}) + (1.7)(350\ \text{kips})$$
$$= 805\ \text{kips}$$

It is required that $\phi P_n \geq P_u$. For axial compression with spiral reinforcement, $\phi = 0.75$. The nominal axial compressive load capacity is given by

$$\begin{aligned} P_n &= 0.85P_o = (0.85)(0.85f'_c A_{\text{concrete}} + f_y A_s) \\ &= (0.85)[0.85f'_c(A_g - A_s) + f_y A_s] \end{aligned}$$

Setting $\phi P_n = P_u$ and solving for the area of longitudinal reinforcing steel gives

$$\begin{aligned} A_s &= \frac{\dfrac{P_u}{\phi(0.85)} - 0.85f'_c A_g}{f_y - 0.85f'_c} \\ &= \frac{\dfrac{(805 \text{ kips})\left(\dfrac{1000 \text{ lbf}}{1 \text{ kip}}\right)}{(0.75)(0.85)} - (0.85)\left(4000 \dfrac{\text{lbf}}{\text{in}^2}\right)(254.5 \text{ in}^2)}{60{,}000 \dfrac{\text{lbf}}{\text{in}^2} - (0.85)\left(4000 \dfrac{\text{lbf}}{\text{in}^2}\right)} \\ &= 7.02 \text{ in}^2 \end{aligned}$$

$A_s = 7.02 \text{ in}^2$ as required for the given applied axial compressive loads is greater than $A_s = 2.54 \text{ in}^2$ as based on the minimum allowed reinforcement ratio, $\rho_g = 0.01$. Therefore, the minimum required area of reinforcement is $A_s = 7.02 \text{ in}^2$.

The round spiral column shown requires six uniformly sized longitudinal reinforcing bars. The required area of each longitudinal reinforcing bar is

$$\begin{aligned} \frac{A_s}{\text{total number of reinforcing bars}} &= \frac{7.02 \text{ in}^2}{6 \text{ bars}} \\ &= 1.17 \text{ in}^2/\text{bar} \end{aligned}$$

A bar area of 1.17 in^2 corresponds to a No. 10 bar, which has a nominal area of 1.27 in^2.

(Note that in an actual design/analysis situation, a check should also be made to see that the actual longitudinal reinforcement ratio does not exceed the maximum allowable ratio of 0.08.)

Answer is D.

38. Determine the amount of reinforcing steel required by the minimum required reinforcement ratio, ρ_g, of 0.01. Since this ratio is defined as $\rho_g = A_s/A_g$, the minimum area of reinforcing steel required is

$$\begin{aligned} A_s &= \rho_g A_g = (0.01)(18 \text{ in})^2 \\ &= 3.24 \text{ in}^2 \end{aligned}$$

Now determine the required amount of reinforcing steel based on the factored axial load, P_u.

$$\begin{aligned} P_u &= 1.4P_{\text{dead}} + 1.7P_{\text{live}} \\ &= (1.4)(150 \text{ kips}) + (1.7)(250 \text{ kips}) \\ &= 635 \text{ kips} \end{aligned}$$

It is required that $\phi P_n \geq P_u$. For axial compression with tied reinforcement, $\phi = 0.70$. The nominal axial compressive load capacity is given by

$$\begin{aligned} P_n &= 0.8P_o = (0.8)(0.85f'_c A_{\text{concrete}} + f_y A_s) \\ &= (0.8)(0.85f'_c(A_g - A_s) + f_y A_s) \end{aligned}$$

Setting $\phi P_n = P_u$ and solving for the area of longitudinal reinforcing steel gives

$$\begin{aligned} A_s &= \frac{\dfrac{P_u}{\phi(0.8)} - 0.85f'_c A_g}{f_y - 0.85f'_c} \\ &= \frac{\dfrac{(635 \text{ kips})\left(\dfrac{1000 \text{ lbf}}{1 \text{ kip}}\right)}{(0.70)(0.8)} - (0.85)\left(4000 \dfrac{\text{lbf}}{\text{in}^2}\right)(324 \text{ in}^2)}{60{,}000 \dfrac{\text{lbf}}{\text{in}^2} - (0.85)\left(4000 \dfrac{\text{lbf}}{\text{in}^2}\right)} \\ &= 0.571 \text{ in}^2 \end{aligned}$$

$A_s = 0.571 \text{ in}^2$ as required for the given applied axial compressive loads is less than $A_s = 3.24 \text{ in}^2$ as based on the minimum allowed reinforcement ratio, $\rho_g = 0.01$. Therefore, the minimum required area of reinforcement is $A_s = 3.24 \text{ in}^2$.

The square tied column shown requires eight uniformly sized longitudinal reinforcing bars. The required area of each longitudinal reinforcing bar is

$$\begin{aligned} \frac{A_s}{\text{total number of reinforcing bars}} &= \frac{3.24 \text{ in}^2}{8 \text{ bars}} \\ &= 0.405 \text{ in}^2/\text{bar} \end{aligned}$$

A bar area of 0.405 in^2 corresponds to a No. 6 bar, which has a nominal area of 0.44 in^2.

(Note that in an actual design and analysis situation, a check should also be made to see that the actual longitudinal reinforcement ratio does not exceed the maximum allowable ratio of 0.08.)

Answer is D.

39. From the "Mechanics of Materials" section of the NCEES Handbook, the design value for the effective column length factors about the x-axis and y-axis, respectively, are $k_x = 0.80$ and $k_y = 0.80$. From the same section in the NCEES Handbook, the modulus of elasticity of steel is

$$E = \left(30 \times 10^6 \ \frac{\text{lbf}}{\text{in}^2}\right)\left(\frac{1 \text{ kip}}{1000 \text{ lbf}}\right) = 30{,}000 \text{ kips/in}^2$$

The unbraced length of the compression member is the same about the x-axis as about the y-axis. Therefore,

$$l_x = l_y = (10 \text{ ft})\left(12 \ \frac{\text{in}}{\text{ft}}\right) = 120 \text{ in}$$

The radius of gyration about the x-axis is

$$r_x = \sqrt{\frac{I_x}{A}} = \sqrt{\frac{533 \text{ in}^4}{19.1 \text{ in}^2}} = 5.28 \text{ in}$$

The slenderness ratio about the x-axis is

$$\text{SR}_x = \frac{k_x l_x}{r_x} = \frac{(0.80)(120 \text{ in})}{5.28 \text{ in}} = 18.2$$

The radius of gyration about the y-axis is

$$r_y = \sqrt{\frac{I_y}{A}} = \sqrt{\frac{174 \text{ in}^4}{19.1 \text{ in}^2}} = 3.02 \text{ in}$$

The slenderness ratio about the y-axis is

$$\text{SR}_y = \frac{k_y l_y}{r_y} = \frac{(0.80)(120 \text{ in})}{3.02 \text{ in}} = 31.8$$

The larger SR controls. Therefore, use $\text{SR} = \text{SR}_y = 31.8$.

The SR is now compared with

$$C_c = \sqrt{\frac{2\pi^2 E}{F_y}} = \sqrt{\frac{(2\pi^2)\left(30{,}000 \ \frac{\text{kips}}{\text{in}^2}\right)}{36 \ \frac{\text{kips}}{\text{in}^2}}} = 128.3$$

Since $\text{SR} = 31.8$ is less than $C_c = 128.3$, the allowable axial compressive stress is given by

$$\begin{aligned} F_a &= \frac{\left(1 - \frac{\left(\frac{kl}{r}\right)^2}{2C_c^2}\right) F_y}{\frac{5}{3} + \frac{(3)\left(\frac{kl}{r}\right)}{8C_c} - \frac{\left(\frac{kl}{r}\right)^3}{8C_c^3}} \\ &= \frac{\left(1 - \frac{(\text{SR})^2}{2C_c^2}\right) F_y}{\frac{5}{3} + \frac{(3)(\text{SR})}{8C_c} - \frac{(\text{SR})^3}{8C_c^3}} \\ &= \frac{\left(1 - \frac{(31.8)^2}{(2)(128.3)^2}\right)\left(36 \ \frac{\text{kips}}{\text{in}^2}\right)}{\frac{5}{3} + \frac{(3)(31.8)}{(8)(128.3)} - \frac{(31.8)^3}{(8)(128.3)^3}} \\ &= 19.85 \text{ kips/in}^2 \end{aligned}$$

Therefore, the total allowable design force is given by

$$P_{\text{total}} = F_a A = \left(19.85 \ \frac{\text{kips}}{\text{in}^2}\right)(19.1 \text{ in}^2) = 379.1 \text{ kips}$$

The allowable design live load can be found by

$$\begin{aligned} P_{\text{total}} &= D + L = P_{\text{dead}} + P_{\text{live}} \\ P_{\text{live}} &= P_{\text{total}} - P_{\text{dead}} = 379.1 \text{ kips} - 7 \text{ kips} \\ &= 372.1 \text{ kips} \quad (370 \text{ kips}) \end{aligned}$$

Answer is B.

40. From the "Mechanics of Materials" section of the NCEES Handbook, the design value for the effective column length factors about the x-axis and y-axis, respectively, are $k_x = 0.80$ and $k_y = 0.80$. From the same section in the NCEES Handbook, the modulus of elasticity of steel is

$$E = \left(30 \times 10^6 \ \frac{\text{lbf}}{\text{in}^2}\right)\left(\frac{1 \text{ kip}}{1000 \text{ lbf}}\right) = 30{,}000 \text{ kips/in}^2$$

The unbraced length of the compression member is the same about the x-axis as about the y-axis. Therefore,

$$l_x = l_y = (10 \text{ ft})\left(12 \ \frac{\text{in}}{\text{ft}}\right) = 120 \text{ in}$$

The radius of gyration about the x-axis is

$$r_x = \sqrt{\frac{I_x}{A}} = \sqrt{\frac{533 \text{ in}^4}{19.1 \text{ in}^2}} = 5.28 \text{ in}$$

The column slenderness parameter about the x-axis is

$$\begin{aligned}\lambda_{cx} &= \left(\frac{k_x l_x}{r_x \pi}\right)\sqrt{\frac{F_y}{E}} \\ &= \left(\frac{(0.80)(120 \text{ in})}{(5.28 \text{ in})\pi}\right)\sqrt{\frac{36 \ \frac{\text{kips}}{\text{in}^2}}{30{,}000 \ \frac{\text{kips}}{\text{in}^2}}} \\ &= 0.20\end{aligned}$$

Since $\lambda_{cx} = 0.20 < 1.5$, the critical compressive stress about the x-axis is

$$\begin{aligned}F_{crx} &= \left(0.658^{\lambda_{cx}^2}\right)F_y = \left(0.658^{(0.20)^2}\right)\left(36 \ \frac{\text{kips}}{\text{in}^2}\right) \\ &= 35.4 \text{ kips/in}^2\end{aligned}$$

The radius of gyration about the y-axis is

$$r_y = \sqrt{\frac{I_y}{A}} = \sqrt{\frac{174 \text{ in}^4}{19.1 \text{ in}^2}} = 3.02 \text{ in}$$

The column slenderness parameter about the y-axis is

$$\begin{aligned}\lambda_{cy} &= \left(\frac{k_y l_y}{r_y \pi}\right)\sqrt{\frac{F_y}{E}} \\ &= \left(\frac{(0.80)(120 \text{ in})}{(3.02 \text{ in})\pi}\right)\sqrt{\frac{36 \ \frac{\text{kips}}{\text{in}^2}}{30{,}000 \ \frac{\text{kips}}{\text{in}^2}}} \\ &= 0.35\end{aligned}$$

Since $\lambda_{cy} = 0.35 < 1.5$, the critical compressive stress about the y-axis is

$$\begin{aligned}F_{cry} &= \left(0.658^{\lambda_{cy}^2}\right)F_y = \left(0.658^{(0.35)^2}\right)\left(36 \ \frac{\text{kips}}{\text{in}^2}\right) \\ &= 34.2 \text{ kips/in}^2\end{aligned}$$

The smaller critical compressive stress controls. Therefore, use

$$F_{cr} = F_{cry} = 34.2 \text{ kips/in}^2$$

The nominal axial compressive strength is

$$P_n = A_g F_{cr} = (19.1 \text{ in}^2)\left(34.2 \ \frac{\text{kips}}{\text{in}^2}\right) = 653.2 \text{ kips}$$

The resistance factor for compression is $\phi_c = 0.85$. The axial compressive design strength is

$$\phi_c P_n = (0.85)(653.2 \text{ kips}) = 555.2 \text{ kips}$$

By setting $P_{\text{total}} = \phi_c P_n = 555.2$ kips, the allowable design live load can be found.

$$P_{\text{total}} = 1.2D + 1.6L = 1.2P_{\text{dead}} + 1.6P_{\text{live}}$$

$$\begin{aligned}P_{\text{live}} &= \frac{P_{\text{total}} - 1.2P_{\text{dead}}}{1.6} = \frac{555.2 \text{ kips} - (1.2)(7 \text{ kips})}{1.6} \\ &= 341.8 \text{ kips} \quad (342 \text{ kips})\end{aligned}$$

Answer is A.

41. It is necessary to check both yielding on the gross area and fracture on the effective net area.

For yielding, the gross area is

$$A_g = (0.5 \text{ in})(2.25 \text{ in} + 3 \text{ in} + 2.25 \text{ in}) = 3.75 \text{ in}^2$$

The allowable tensile stress for yielding is

$$F_t = 0.6F_y = (0.6)\left(36 \ \frac{\text{kips}}{\text{in}^2}\right) = 21.6 \text{ kips/in}^2$$

The allowable tensile force for yielding is

$$P_{\text{total}} = F_t A_g = \left(21.6 \ \frac{\text{kips}}{\text{in}^2}\right)(3.75 \text{ in}^2) = 81.0 \text{ kips}$$

For fracture, in order to determine the net area, the controlling net width of the member must be determined.

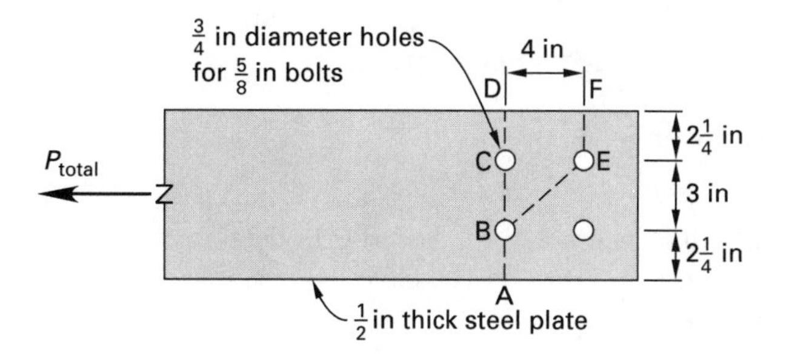

For line ABCD across the member width, the net width is

$$\begin{aligned}b_n &= b - \sum d \\ &= (2.25 \text{ in} + 3 \text{ in} + 2.25 \text{ in}) - (2 \text{ holes})(0.75 \text{ in}) \\ &= 6 \text{ in}\end{aligned}$$

For line ABEF across the member width, the net width is

$$b_n = b - \sum d + \frac{\sum s^2}{4g}$$
$$= (2.25\ \text{in} + 3\ \text{in} + 2.25\ \text{in}) - (2\ \text{holes})(0.75\ \text{in}) + \frac{(1\ \text{space})(4\ \text{in})^2}{(4)(3\ \text{in})}$$
$$= 7.33\ \text{in}$$

The smaller net width controls. Therefore, use $b_n = 6$ in.

The net area is

$$A_n = b_n t = (6\ \text{in})(0.5\ \text{in}) = 3\ \text{in}^2$$

Since there are two fasteners (bolts) per line, $U = 0.75$ and the effective net area is

$$A_e = UA_n = (0.75)(3\ \text{in}^2) = 2.25\ \text{in}^2$$

The allowable tensile stress for fracture is

$$F_t = 0.5F_u = (0.5)\left(58\ \frac{\text{kips}}{\text{in}^2}\right) = 29\ \text{kips/in}^2$$

The allowable tensile force for fracture is

$$P_{\text{total}} = F_t A_e = \left(29\ \frac{\text{kips}}{\text{in}^2}\right)(2.25\ \text{in}^2) = 65.2\ \text{kips}$$

$P_{\text{total}} = 65.2$ kips for fracture is smaller than $P_{\text{total}} = 81.0$ kips for yielding, so use $P_{\text{total}} = 65.2$ kips. The allowable design live load can be found by

$$P_{\text{total}} = D + L = P_{\text{dead}} + P_{\text{live}}$$
$$P_{\text{live}} = P_{\text{total}} - P_{\text{dead}} = 65.2\ \text{kips} - 15\ \text{kips}$$
$$= 50.2\ \text{kips}\quad (50\ \text{kips})$$

Answer is A.

42. It is necessary to check both yielding on the gross area and fracture on the effective net area.

For yielding, the gross area is

$$A_g = (0.5\ \text{in})(2.25\ \text{in} + 3\ \text{in} + 2.25\ \text{in}) = 3.75\ \text{in}^2$$

The allowable tensile force for yielding is

$$\phi_t P_n = 0.9F_y A_g = (0.9)\left(36\ \frac{\text{kips}}{\text{in}^2}\right)(3.75\ \text{in}^2)$$
$$= 121.5\ \text{kips}$$

For fracture, in order to determine the net area, the controlling net width of the member must be determined. From Prob. 41, the effective net area is $A_e = 2.25\ \text{in}^2$.

The allowable tensile force for fracture is

$$\phi_t P_n = 0.75F_u A_e = (0.75)\left(58\ \frac{\text{kips}}{\text{in}^2}\right)(2.25\ \text{in}^2)$$
$$= 97.9\ \text{kips}$$

$\phi_t P_n = 97.9$ kips for fracture is smaller than $\phi_t P_n = 121.5$ kips for yielding, so use $\phi_t P_n = 97.9$ kips.

By setting $P_{\text{total}} = \phi_t P_n = 97.9$ kips, the allowable design live load can be found by

$$P_{\text{total}} = 1.2D + 1.6L = 1.2P_{\text{dead}} + 1.6P_{\text{live}}$$
$$P_{\text{live}} = \frac{P_{\text{total}} - 1.2P_{\text{dead}}}{1.6} = \frac{97.9\ \text{kips} - (1.2)(15\ \text{kips})}{1.6}$$
$$= 49.9\ \text{kips}\quad (50\ \text{kips})$$

Answer is A.

43. By Simpson's 1/3 rule,

$$\text{area} = (w)\left(\frac{h_1 + (2)\left(\sum h_{\text{odds}}\right) + (4)\left(\sum h_{\text{evens}}\right) + h_5}{3}\right)$$
$$= (5\ \text{m})\left(\frac{4.2\ \text{m} + (2)(6.7\ \text{m}) + (4)(8.1\ \text{m} + 7.6\ \text{m}) + 8.3\ \text{m}}{3}\right)$$
$$= 147.8\ \text{m}^2\quad (148\ \text{m}^2)$$

Answer is C.

44. From the PVC at sta 10+35 to the location of desired elevation at sta 11+35, $x = 1135\ \text{m} - 1035\ \text{m} = 100$ m. The horizontal vertical curve length, from PVC to PVT, is given in the problem statement as $L = 350$ m. The grades are given as $g_1 = +7\%$ and $g_2 = -5\%$. The elevation of the PVC is given as $y_{\text{PVC}} = 60.0$ m.

Therefore, the curve elevation at sta 11+35 is

$$\begin{aligned} y_{11+35} &= y_{\text{PVC}} + g_1 x + \left(\frac{g_2 - g_1}{2L}\right) x^2 \\ &= 60.0 \text{ m} + (+7\%)\left(\frac{1}{100\%}\right)(100 \text{ m}) \\ &\quad + \left(\frac{(-5\% - (+7\%))\left(\frac{1}{100\%}\right)}{(2)(350 \text{ m})}\right)(100 \text{ m})^2 \\ &= 65.3 \text{ m} \quad (65 \text{ m}) \end{aligned}$$

Answer is A.

45. The PVI is located at $x = L/2 = (350 \text{ m})/2 = 175$ m from the PVC. From the tangent elevation equation, the tangent elevation is found to be

$$\begin{aligned} y_{\text{PVC}} + g_1 x &= 60.0 \text{ m} + (+7\%)\left(\frac{1}{100\%}\right)(175 \text{ m}) \\ &= 72.2 \text{ m} \quad (72 \text{ m}) \end{aligned}$$

Answer is D.

46. The 4 m long rod can be described as extending from the center of a 4 m radius circle. The end of the rod is at the top of the circle when it is truly vertical and makes a circular arc when it goes out of plumb.

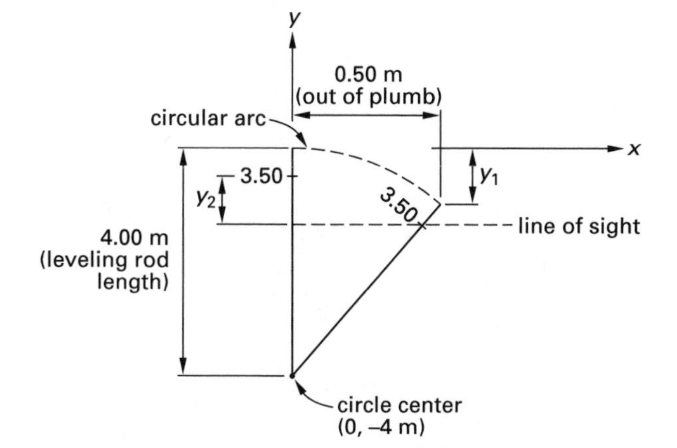

With the reference taken at the top of the circle, the center is at coordinates $h = x = 0$ ft and $k = y = -4$ m. The equation of a circle with the center at (h, k) and a radius, r, is

$$\begin{aligned} (x - h)^2 + (y - k)^2 &= r^2 \\ (x - 0 \text{ m})^2 + (y + 4 \text{ m})^2 &= (4 \text{ m})^2 \end{aligned}$$

Removing units for simplicity (but remembering that all distances are in meters) and simplifying the equation gives

$$y^2 + 8y + x^2 = 0$$

From this, the change in vertical distance at the top end of the rod when it goes 0.50 m out of plumb can be solved by letting $x = 0.50$ and solving for y.

$$\begin{aligned} y^2 + 8y + x^2 &= 0 \\ y^2 + 8y + (0.50)^2 &= 0 \\ y^2 + 8y + 0.25 &= 0 \\ y &= -0.0314 \quad \text{[the nontrivial solution]} \end{aligned}$$

Therefore, the change in vertical distance of the leveling rod end is $y_1 = -y = 0.0314$ m.

A ratio of the smaller circular arc with a radius of 3.50 m to the larger circular arc with a radius of 4.00 m can be used to find y_2.

$$\frac{y_2}{3.50 \text{ m}} = \frac{y_1}{4.00 \text{ m}}$$

$$\begin{aligned} y_2 &= (3.50 \text{ m})\left(\frac{y_1}{4.00 \text{ m}}\right) = (3.50 \text{ m})\left(\frac{0.0314 \text{ m}}{4.00 \text{ m}}\right) \\ &= 0.0275 \text{ m} \end{aligned}$$

Since the out-of-plumb reading was 3.50 m, the correct reading with the rod truly vertical is

$$3.50 \text{ m} - y_2 = 3.50 \text{ m} - 0.0275 \text{ m} = 3.47 \text{ m}$$

Answer is B.

47. A sketch of the required distances is helpful.

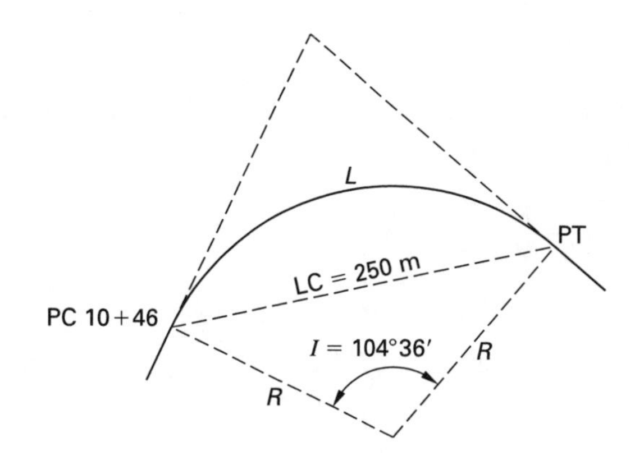

The intersection of angle, I, should be converted into a decimal angle to give

$$I = 104° + (36')\left(\frac{1°}{60'}\right) = 104.6°$$

The radius of the curve is

$$R = \frac{\text{LC}}{2\sin\left(\frac{I}{2}\right)} = \frac{250\text{ m}}{2\sin\left(\frac{104.6^\circ}{2}\right)} = 158.0\text{ m}$$

Answer is A.

48. From Prob. 47, $R = 158.0$ m. The length of curve from PC to PT is

$$L = RI = (158.0\text{ m})(104.6^\circ)\left(\frac{\pi\text{ rad}}{180^\circ}\right) = 288.4\text{ m}$$

Therefore, adding this to the PC station, the PT station is at 1046.0 m + 288.4 m = 1334.4 m = sta 13+34.4.

Answer is C.

49. The degree of curvature is converted into radians and the radius of curvature is found to be

$$R = \frac{s}{\phi} = \frac{700\text{ ft}}{(10^\circ)\left(\frac{\pi\text{ rad}}{180^\circ}\right)} = 4010.7\text{ ft}$$

Therefore, the minimum required length of spiral transition onto this road is

$$L_s = (1.6)\left(\frac{V^3}{R}\right) = (1.6)\left(\frac{\left(45\ \frac{\text{mi}}{\text{hr}}\right)^3}{4010.7\text{ ft}}\right)$$

$$= 36.4\text{ ft}\quad(36\text{ ft})$$

Answer is B.

50. A vehicle will require a longer braking distance if it is traveling downhill. Therefore, S has to be subtracted from the denominator to result in the largest design braking distance. The braking distance is found to be

$$d = \frac{\text{v}^2}{2g(f \pm S)} + t\text{v}$$

$$= \frac{\left[\left(80\ \frac{\text{km}}{\text{h}}\right)\left(\frac{1000\text{ m}}{1\text{ km}}\right)\left(\frac{1\text{ h}}{3600\text{ s}}\right)\right]^2}{(2)\left(9.81\ \frac{\text{m}}{\text{s}^2}\right)\left(0.35 - \frac{1.5\%}{100}\right)}$$

$$+ (2.0\text{ s})\left(80\ \frac{\text{km}}{\text{h}}\right)\left(\frac{1000\text{ m}}{1\text{ km}}\right)\left(\frac{1\text{ h}}{3600\text{ s}}\right)$$

$$= 119.6\text{ m}\quad(120\text{ m})$$

Answer is D.

51. There are two equations to check.

For $S < L$,

$$L = \frac{AS^2}{(100)\left(\sqrt{2H_1} + \sqrt{2H_2}\right)^2}$$

$$= \frac{\left(1\% - (-3\%)\right)(300\text{ m})^2}{(100)\left(\sqrt{(2)(1.2\text{ m})} + \sqrt{(2)(0.2\text{ m})}\right)^2}$$

$$= 756\text{ m}$$

For $S > L$,

$$L = 2S - \frac{(200)\left(\sqrt{H_1} + \sqrt{H_2}\right)^2}{A}$$

$$= (2)(300\text{ m}) - \frac{(200)\left(\sqrt{1.2\text{ m}} + \sqrt{0.2\text{ m}}\right)^2}{1\% - (-3\%)}$$

$$= 481\text{ m}$$

Since the previous two equations show that $S < L$, use $L = 756$ m (760 m) as the minimum required vertical curve length.

Answer is C.

52. The best runway orientation for this weight of aircraft minimizes the frequency of crosswinds with speeds greater than 15 mi/hr. This can be determined by plotting a wind rose where the percentages of winds are placed in the proper wind direction/speed locations.

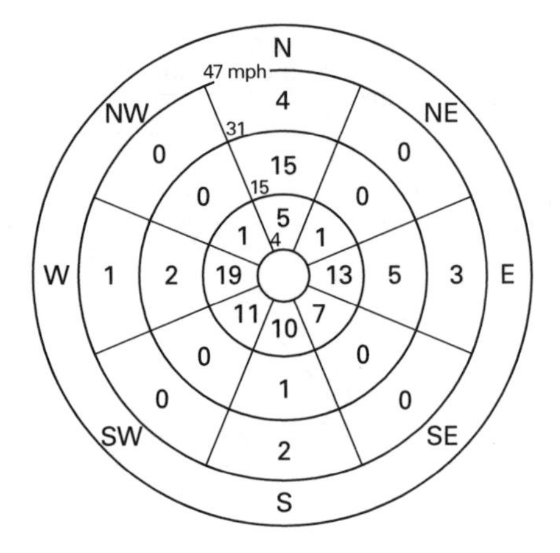

For aircraft heavier than 12,500 lbf in total weight, the crosswind component is 15 mph maximum. Therefore, parallel lines tangent to the 15 mph circle on the wind rose are drawn and rotated, with the centerline in a chosen orientation, to determine total percentages in that orientation. The total percentages are determined by adding up all percentages and fractions of percentages that are enclosed by the parallel lines. These total percentages tell how often a runway with a certain orientation may be used with crosswinds equal to or less than 15 mph.

For the N-S orientation,

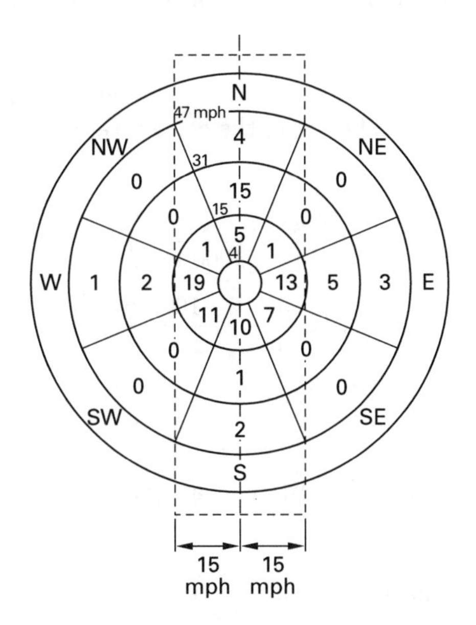

$$
\begin{aligned}
\text{N-S\%} &= 1\% + 5\% + 1\% + 13\% + 7\% + 10\% + 11\% \\
&\quad + 19\% + 15\% + 1\% + 4\% + 2\% \\
&= 89\%
\end{aligned}
$$

For the NW-SE orientation,

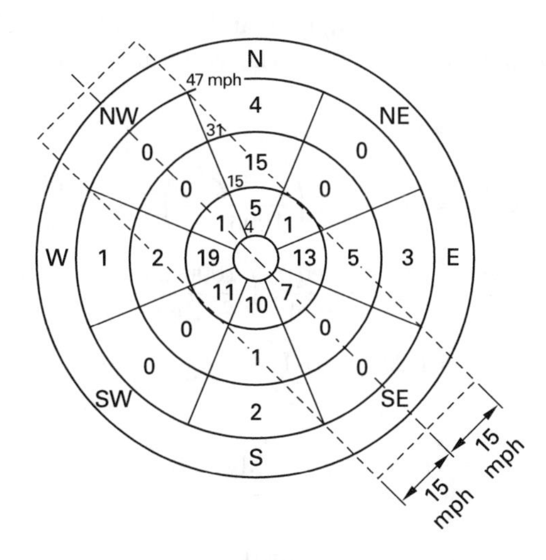

$$
\begin{aligned}
\text{NW-SE\%} &= 1\% + 5\% + 1\% + 13\% + 7\% + 10\% \\
&\quad + 11\% + 19\% + (0.5)(2\%) + (0.5)(15\%) \\
&\quad + (0.5)(1\%) + (0.5)(5\%) + (0.1)(1\%) \\
&\quad + (0.1)(4\%) + (0.1)(2\%) + (0.1)(3\%) \\
&= 79.5\%
\end{aligned}
$$

For the SW-NE orientation,

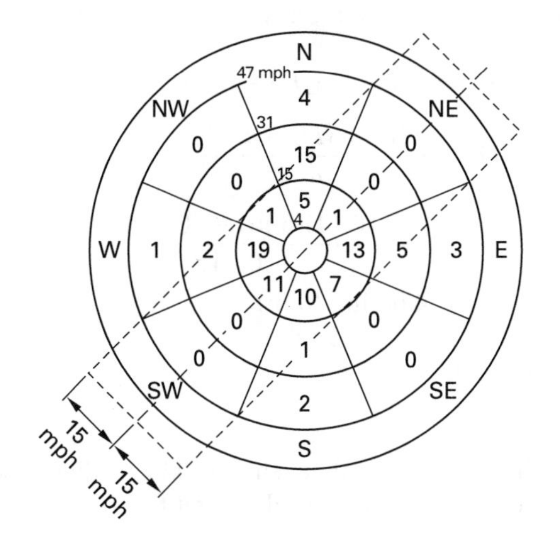

$$
\begin{aligned}
\text{SW-NE\%} &= 1\% + 5\% + 1\% + 13\% + 7\% + 10\% \\
&\quad + 11\% + 19\% + (0.5)(15\%) + (0.5)(5\%) \\
&\quad + (0.5)(2\%) + (0.5)(1\%) + (0.1)(4\%) \\
&\quad + (0.1)(3\%) + (0.1)(1\%) + (0.1)(2\%) \\
&= 79.5\%
\end{aligned}
$$

For the E-W orientation,

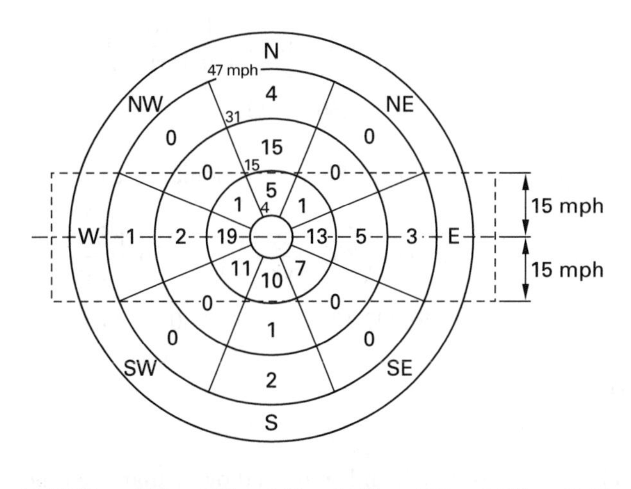

$$
\begin{aligned}
\text{E-W\%} &= 1\% + 5\% + 1\% + 13\% + 7\% + 10\% + 11\% \\
&\quad + 19\% + 5\% + 2\% + 3\% + 1\% \\
&= 78\%
\end{aligned}
$$

(Note that for some orientations, the parallel lines cover negligibly small parts of some percentages in the wind rose that can be disregarded.)

The N-S orientation shows the largest percentage, 89%, that the runway may be used with crosswinds equal to or less than 15 mph, which indicates that this is the best runway orientation.

Answer is A.

53. The AASHTO structural number equation can be rearranged to solve for surface thickness, D_1, to give

$$\mathrm{SN} = a_1 D_1 + a_2 D_2 + a_3 D_3$$

$$\begin{aligned} D_1 &= \frac{\mathrm{SN} - a_2 D_2 - a_3 D_3}{a_1} \\ &= \frac{4 - (0.14)(6 \text{ in}) - (0.11)(10 \text{ in})}{0.44} \\ &= 4.68 \text{ in} \quad (5 \text{ in}) \end{aligned}$$

Answer is C.

54. The total equivalent single axle loading (ESAL) per truck for each trip is calculated to be

$$\begin{aligned} \mathrm{ESAL_{truck}} &= (1 \text{ single axle})(0.0877) \\ &\quad + (2 \text{ tandem axles})(0.1206) \\ &= 0.3289 \text{ ESAL/truck-trip} \end{aligned}$$

The total daily ESAL for 40 trucks, each making 10 trips a day, is calculated to be

$$\begin{aligned} \mathrm{ESAL_{day}} &= (40 \text{ trucks})\left(10\ \frac{\text{trips}}{\text{day}}\right) \\ &\quad \times \left(0.3289\ \frac{\text{ESAL}}{\text{truck-trip}}\right) \\ &= 131.56 \text{ ESAL/day} \end{aligned}$$

For five years, the total ESAL is calculated to be

$$\begin{aligned} \mathrm{ESAL_{5\ yr}} &= (5 \text{ yr})\left(365\ \frac{\text{days}}{\text{yr}}\right)\left(131.56\ \frac{\text{ESAL}}{\text{day}}\right) \\ &= 240{,}097 \text{ ESAL} \quad (240{,}000 \text{ ESAL}) \end{aligned}$$

Answer is D.

55. The amount of solids removed by primary sedimentation is

$$(0.65)\left(350\ \frac{\text{mg}}{\text{L}}\right)\left(\frac{1 \text{ g}}{1000 \text{ mg}}\right) = 0.2275 \text{ g/L}$$

For 5% solids in the resulting sludge, the total amount of sludge produced per liter of wastewater, x, is

$$\frac{0.2275 \text{ g}}{x} = 0.05$$

$$x = 4.55 \text{ g}$$

Answer is D.

56. Water hardness is caused by multivalent cations, especially the divalent calcium, magnesium, strontium, iron, and manganese cations. Accordingly, only Ca^{2+} and Mg^{2+} cause water hardness in the given water sample analysis. Equivalent and molecular weight information for these two cations can be found in the table in the "Civil Engineering" section in the NCEES Handbook.

For Ca^{2+}, n = 2 eq/mol and molecular weight = 40.078 g/mol. Therefore, the concentration of Ca^{2+} is

$$\begin{aligned} [\mathrm{Ca^{2+}}] &= \left(10\ \frac{\text{mg}}{\text{L}}\right)\left(\frac{1 \text{ g}}{1000 \text{ mg}}\right)\left(\frac{1 \text{ mol}}{40.078 \text{ g}}\right) \\ &\quad \times \left(2\ \frac{\text{eq}}{\text{mol}}\right)\left(\frac{1000 \text{ meq}}{1 \text{ eq}}\right) \\ &= 0.499 \text{ meq/L} \end{aligned}$$

The hardness caused by the Ca^{2+} is

$$\begin{aligned} \mathrm{Ca^{2+}} \text{ hardness} &= \left(0.499\ \frac{\text{meq}}{\text{L}}\right) \\ &\quad \times \left(\frac{50\ \frac{\text{mg}}{\text{L}} \text{ as } \mathrm{CaCO_3} \text{ equivalents}}{1\ \frac{\text{meq}}{\text{L}}}\right) \\ &= 24.95 \text{ mg/L as } \mathrm{CaCO_3} \text{ equivalents} \end{aligned}$$

For Mg^{2+}, n = 2 eq/mol and molecular weight = 24.305 g/mol. Therefore, the concentration of Mg^{2+} is

$$\begin{aligned} [\mathrm{Mg^{2+}}] &= \left(15\ \frac{\text{mg}}{\text{L}}\right)\left(\frac{1 \text{ g}}{1000 \text{ mg}}\right)\left(\frac{1 \text{ mol}}{24.305 \text{ g}}\right) \\ &\quad \times \left(2\ \frac{\text{eq}}{\text{mol}}\right)\left(\frac{1000 \text{ meq}}{1 \text{ eq}}\right) \\ &= 1.234 \text{ meq/L} \end{aligned}$$

The hardness caused by the Mg^{2+} is

$$\begin{aligned} \mathrm{Mg^{2+}} \text{ hardness} &= \left(1.234\ \frac{\text{meq}}{\text{L}}\right) \\ &\quad \times \left(\frac{50\ \frac{\text{mg}}{\text{L}} \text{ as } \mathrm{CaCO_3} \text{ equivalents}}{1\ \frac{\text{meq}}{\text{L}}}\right) \\ &= 61.7 \text{ mg/L as } \mathrm{CaCO_3} \text{ equivalents} \end{aligned}$$

The total water hardness, as caused by both the Ca^{2+} and Mg^{2+} cations, is the sum of the two individual hardnesses.

$$\begin{aligned}\text{total hardness} &= Ca^{2+}\text{ hardness} + Mg^{2+}\text{ hardness}\\ &= 24.95\ \frac{\text{mg}}{\text{L}}\text{ as CaCO}_3\text{ equivalents}\\ &\quad + 61.7\ \frac{\text{mg}}{\text{L}}\text{ as CaCO}_3\text{ equivalents}\\ &= 86.6\text{ mg/L as CaCO}_3\text{ equivalents}\\ &\quad (87\text{ mg/L as CaCO}_3\text{ equivalents})\end{aligned}$$

Answer is C.

57. The cross-sectional area of the channel is

$$A_x = (\text{width})(\text{depth}) = (1.5\text{ m})(3\text{ m}) = 4.5\text{ m}^2$$

Therefore, the approach velocity is

$$\frac{Q}{A_x} = \frac{0.5\ \frac{\text{m}^3}{\text{s}}}{4.5\text{ m}^2} = 0.1\text{ m/s}$$

Answer is A.

58. The total water hardness in this problem is caused by magnesium carbonate hardness. Therefore, to reduce the water hardness to the desired level from 75 mg/L as $CaCO_3$ equivalents to 5 mg/L as $CaCO_3$ equivalents, the amount of hardness that must be eliminated is

$$\begin{aligned}&75\ \frac{\text{mg}}{\text{L}}\text{ as CaCO}_3\text{ equivalents}\\ &\quad - 5\ \frac{\text{mg}}{\text{L}}\text{ as CaCO}_3\text{ equivalents}\\ &\quad = 70\text{ mg/L as CaCO}_3\text{ equivalents}\end{aligned}$$

Conversion from mg/L as $CaCO_3$ equivalents requires use of the equivalent and molecular weight information along with the lime-soda softening chemical equations given in the "Civil Engineering" section in the NCEES Handbook. For magnesium carbonate hardness removal, the appropriate lime-soda softening chemical equation is

$$\begin{aligned}&Mg(HCO_3)_2 + 2Ca(OH)_2 \longrightarrow\\ &\quad 2CaCO_3(s) + Mg(OH)_2(s) + 2H_2O\end{aligned}$$

From the NCEES Handbook, $n = 2$ eq/mol for $Mg(HCO_3)_2$. The concentration of magnesium carbonate hardness that must be removed is

$$\begin{aligned}&\left[\frac{\left(70\ \frac{\text{mg}}{\text{L}}\text{ as CaCO}_3\text{ equivalents}\right)}{\left(\frac{50\ \frac{\text{mg}}{\text{L}}\text{ as CaCO}_3\text{ equivalents}}{1\ \frac{\text{meq}}{\text{L}}\ Mg(HCO_3)_2}\right)}\right]\\ &\quad\times\left(\frac{1\text{ eq}}{1000\text{ meq}}\right)\left(\frac{1}{n}\right)\\ &= \left[\frac{\left(70\ \frac{\text{mg}}{\text{L}}\text{ as CaCO}_3\text{ equivalents}\right)}{\left(\frac{50\ \frac{\text{mg}}{\text{L}}\text{ as CaCO}_3\text{ equivalents}}{1\ \frac{\text{meq}}{\text{L}}\ Mg(HCO_3)_2}\right)}\right]\\ &\quad\times\left(\frac{1\text{ eq }Mg(HCO_3)_2}{1000\text{ meq }Mg(HCO_3)_2}\right)\\ &\quad\times\left(\frac{1\text{ eq }Mg(HCO_3)_2}{2\text{ mol }Mg(HCO_3)_2}\right)\\ &= 0.0007\text{ mol/L }Mg(HCO_3)_2\end{aligned}$$

From the chemical equation, removal of 1 mol of magnesium carbonate requires 2 mol of calcium hydroxide. Therefore, the concentration of calcium hydroxide that must be added is

$$\begin{aligned}[Ca(OH)_2] &= \left(2\ \frac{\text{mol }Ca(OH)_2}{\text{mol }Mg(HCO_3)_2}\right)\\ &\quad\times\left(0.0007\ \frac{\text{mol}}{\text{L}}\ Mg(HCO_3)_2\right)\\ &= 0.0014\text{ mol/L }Ca(OH)_2\\ &\quad (0.0014\text{ M }Ca(OH)_2)\end{aligned}$$

Answer is C.

59. From the given graph, for 35% BOD removal the surface overflow rate is 50 m/day. The total cross-sectional area required for both basins is

$$A = \frac{16\,000\ \frac{\text{m}^3}{\text{day}}}{50\ \frac{\text{m}}{\text{day}}} = 320\text{ m}^2$$

The volume per unit length of basin, based on area, A, and depth, D, is

$$\begin{aligned}V = AL = DWL &= (320\text{ m}^2)(1\text{ m}) = 320\text{ m}^3\\ &= D(320\text{ m}^2)\end{aligned}$$

From the NCEES Handbook equation for hydraulic residence time and based on a unit length of basin, the relationship required for depth, D, can be determined.

$$\text{hydraulic residence time} = \frac{V}{Q}$$

$$V = (Q)(\text{hydraulic residence time}) = (D)(320\ \text{m}^2)$$

$$\begin{aligned} D &= \frac{(Q)(\text{hydraulic residence time})}{320\ \text{m}^2} \\ &= \frac{\left(16\,000\ \frac{\text{m}^3}{\text{day}}\right)\left(\frac{1\ \text{day}}{24\ \text{h}}\right)(1\ \text{h})}{320\ \text{m}^2} \\ &= 2.08\ \text{m} \quad (2.1\ \text{m}) \end{aligned}$$

Answer is D.

60.

$$\begin{aligned} A &= \text{total surface area} = LW \\ &= (2\ \text{basins})(20\ \text{m})\left(5\ \frac{\text{m}}{\text{basin}}\right) \\ &= 200\ \text{m}^2 \end{aligned}$$

The hydraulic loading rate is

$$\begin{aligned} \text{hydraulic loading rate} &= \frac{Q}{A} \\ &= \frac{\left(16\,000\ \frac{\text{m}^3}{\text{day}}\right)\left(\frac{1\ \text{day}}{24\ \text{h}}\right)}{200\ \text{m}^2} \\ &= 3.3\ \text{m}^3/\text{m}^2{\cdot}\text{h} \end{aligned}$$

Answer is D.

More FE/EIT Exam Practice!

All FE/EIT examinees take the same General test in the morning session of the FE exam. To prepare for this test, you can use any of the following publications, depending on how much and what kind of review you need.

FE Review Manual: Rapid Preparation for the General Fundamentals of Engineering Exam

Michael R. Lindeburg, PE

The *FE Review Manual* prepares you for the FE/EIT exam in the fastest, most efficient way possible. Completely up-to-date for the exam content and format, this book gives you diagnostic tests to see what you need to study most, concise reviews of all exam topics, 1150+ practice problems with solutions, and a complete eight-hour sample exam with solutions. The *FE Review Manual* is your best choice if you are still in college or a recent graduate, or if your study time is limited.

999 Nonquantitative Problems for FE Examination Review

Kenton Whitehead, PhD, PE

Passing the FE exam means answering questions quickly and accurately. Nonquantitative problems on the exam don't require numerical calculations but rather an understanding of theory and principle. It's essential that you answer these questions quickly, leaving yourself more time to work on quantitative problems. This book will bring you up to speed on the concepts you need to know and give you intensive practice with problems you won't find anywhere else. Answers are included.

FE/EIT Sample Examinations

Michael R. Lindeburg, PE

Testing yourself with a realistic simulation of the FE exam is an essential part of your preparation. There's no better way to get ready for the time pressure of the exam. *FE/EIT Sample Examinations* includes two complete eight-hour practice tests in multiple-choice format. Fully worked-out solutions are included. Use these sample exams to build skills and confidence for taking the general FE exam.

Calculus Refresher for the Fundamentals of Engineering Exam

Peter Schiavone, PhD

Many engineers report having more trouble with problems involving calculus than with anything else on the FE/EIT exam. This book covers all the areas that you'll need to know for the exam: differential and integral calculus, centroids and moments of inertia, differential equations, and precalculus topics such as quadratic equations and trigonometry. You get clear explanations of theory, relevant examples, and FE-style practice problems (with solutions). If you are at all unsure of your calculus skills, this book is a must.

Professional Publications, Inc.
1250 Fifth Avenue • Belmont, CA 94002
(800) 426-1178 • Fax (650) 592-4519
For secure, convenient ordering, try our complete online catalog at
www.ppi2pass.com

Reporting an error . . .

BEFORE reporting this error, please take a moment to check the errata listing on the PPI website at **www.ppi2pass.com/Errata/Errata.html**. The error(s) you noticed may have already been identified.

Title of this book: ____________________ Edition: ________ Printing: ________

The error occurs on page ____________. Please describe the error and how you think it should be corrected:

Name ____________________

Address ____________________

City/State/Zip ____________________

Phone ____________________ Email ____________________

PROFESSIONAL PUBLICATIONS, INC.
800-426-1178 • Fax 650-592-4519
errata@ppi2pass.com

NO POSTAGE
NECESSARY
IF MAILED
IN THE
UNITED STATES

POSTAGE WILL BE PAID BY ADDRESSEE

PROFESSIONAL PUBLICATIONS INC
1250 FIFTH AVE
BELMONT CA 94002-9979

NO POSTAGE
NECESSARY
IF MAILED
IN THE
UNITED STATES

BUSINESS REPLY MAIL
FIRST CLASS MAIL PERMIT NO. 33 BELMONT, CA

POSTAGE WILL BE PAID BY ADDRESSEE

PROFESSIONAL PUBLICATIONS INC
1250 FIFTH AVE
BELMONT CA 94002-9979